Applied Statistics
A Handbook of GENSTAT Analyses

OTHER STATISTICS TEXTS FROM
CHAPMAN AND HALL

Further information of the complete range of Chapman and Hall *statistics books is available from the publishers.*

Applied Statistics

A Handbook of GENSTAT Analyses

E. J. Snell

Department of Mathematics, Imperial College, London

and

H. R. Simpson

formerly at the Statistics Department, Rothamsted

A complement to
Applied Statistics
Principles and Examples
D. R. Cox and E. J. Snell

CHAPMAN & HALL
London · New York · Tokyo · Melbourne · Madras

UK   Chapman & Hall, 2–6 Boundary Row, London SE1 8HN

USA   Chapman & Hall, 29 West 35th Street, New York NY10001

JAPAN   Chapman & Hall Japan, Thomson Publishing Japan,
            Hirakawacho Nemoto Building, 7F, 1-7-11 Hirakawa-cho,
            Chiyoda-ku, Tokyo 102

AUSTRALIA   Chapman & Hall Australia, Thomas Nelson Australia,
            102 Dodds Street, South Melbourne, Victoria 3205

INDIA   Chapman & Hall India, R. Seshadri, 32 Second Main Road,
            CIT East, Madras 600 035

First edition 1991

© 1991 E. J. Snell and H. R. Simpson

Typeset in 10/12 Times by
KEYTEC, Bridport, Dorset
Printed in Great Britain by
T. J. Press (Padstow) Ltd, Padstow, Cornwall

ISBN 0 412 35320 2

British Library Cataloguing in Publication Data
Snell, E. J.
    Applied statistics.
    1. Statistical analysis. Applications of computer
    systems. Software packages. GENSTAT
    I. Title II. Simpson, H. R.
    519.502855369

    ISBN 0–412–35320–2

Library of Congress Cataloging-in-Publication Data
Snell, E. J.
    Applied statistics: a handbook of Genstat analyses/
    E. J. Snell, H. R. Simpson. – 1st ed.
        p.   cm.
    Includes bibliographical references and index.
    ISBN 0–412–35320–2
    1. Genstat (Computer system) 2. Mathematical
    statistics–Data processing. 3. Statistics–Data processing. I.
    Simpson, H. R. (Howard R.) 1931– .II. Title.
        QA276.4.S62 1991
    519.5'0285'5369–dc20   90-20979
                                                        CIP

Contents

Preface

Genstat is a general computer program for statistical analysis. Its great flexibility is demonstrated in this handbook by analysing the wide range of examples discussed in *Applied Statistics. Principles and Examples* (Cox and Snell, 1981). Attention there concentrates not on computational procedures but on general statistical methodology and the interpretation of conclusions. Most of the data sets are small and sometimes capable of analysis by hand but often it is these seemingly small problems which involve the most tricky computational procedures. The flexibility of Genstat copes with all the examples of *Applied Statistics*.

The Genstat programs listed here have been written by H. R. Simpson.

We are indebted to J. C. Gower for suggesting the writing of this handbook and for his encouragement throughout its preparation; to members of the Genstat 5 Committee for their help and to Rothamsted Experimental Station for the computing facilities extended to H. R. S., a visiting worker there following his retirement from their Statistics Department.

E. J. Snell
H. R. Simpson

PREFACE

Foreword

Genstat was originally written to enable Rothamsted statisticians to do their everyday work with the minimum of fuss. Drawing on experience of using computers from 1954 onwards, the key to a successful system was seen to lie in flexibility and unification. Flexibility is needed to describe the enormous number of ways in which data are collected, for presenting results in readable forms (both numerical and graphical) and for describing the operations needed for statistical work. Flexibility is achieved through a powerful underlying programming language which operates on a general class of data-structures to produce new structures from the same class. Unification of statistical methodology is achieved in various ways: John Nelder's unification of the theory of Generalised Linear Models (GLMs) and his method for describing linear models used both with GLMs and to describe designed experiments, Graham Wilkinson's general recursive algorithm for analysis of variance, the matrix algebra which binds together much of multivariate analysis, function optimization that underlies non-linear model fitting by least-squares and maximum likelihood, and the Box-Jenkins approach to time-series analysis. Genstat does not aim to be exhaustive but it has a procedure mechanism (and a growing standard Procedure Library) that allows unlimited extension in whatever direction users may find useful. Yet because of these unifications the total computer code of Genstat is quite modest. Flexibility extends to the use of graphical interfaces, thanks to the Graphical Supplement of the Numerical Algorithms Group, Oxford, and attention to portability considerations that allows Genstat to be transferred readily to a wide range of computers, including PCs and workstations.

Howard Simpson has been deeply involved with this continuing development since its inception. Joyce Snell is, of course, a co-author of the very successful *Applied Statistics. Principles and Examples* (Cox and Snell, 1981). This handbook will show readers what Genstat can do and will, I hope, dispel some popular misconceptions. Genstat is not a statistical package (although it has some package-like features and can be so used for much routine work); Genstat is not primarily an analysis of variance program (although it has an excellent ANOVA facility); Genstat is not only for the expert statistician (although it provides flexible facilities invaluable to the research statistician); Genstat is not a batch-mode program (although it can be so used, but interactive working is the norm for exploratory work and for program development). Rather, Genstat is all these things and can be used at many levels. Read on and see for yourself.

John Gower
Leiden, The Netherlands

Introduction

Genstat programs are provided for each of the examples discussed in *Applied Statistics. Principles and Examples* (Cox and Snell, 1981), hereafter referred to simply as *App. Stat.*

Genstat is a general purpose statistical computing system with an extremely flexible command language operating on a variety of data structures. It may be used on a number of computer ranges, either interactively for exploratory analysis or in batch mode for standard analysis.

The aim throughout this handbook is to demonstrate the flexibility of Genstat and in particular to show how to reproduce the tables and results discussed in *App. Stat.* The examples are labelled Examples A, ..., X as in *App. Stat.* and tables of results etc. given there are referred to, for example, as Table A.2, *App. Stat.*

Each set of data is reproduced in full, with a description of the data, exactly as in *App. Stat.* in order to make this handbook, to some extent, self-contained although it must be used in parallel with *App. Stat.* if full benefit is to be gained from an analysis.

Genstat programs are listed for each example together with guiding comments enclosed in quotes (".. .") as additional help for the reader. The statistical analyses range from the very simple, barely justifying the use of a computer, to the complex but for the sake of completeness each Example A, ..., X is included.

A common difficulty for any beginner presented with a new computer package is one of simply getting started. A beginner will find the following section, **An introduction to the Genstat statistical system**, helpful.

In practice one of the best ways to learn to write a program for a particular problem is to follow one already written for a similar example and to refer to the *Genstat manual* (OUP, 1987) for further details as necessary. The wide range of applications covered in this handbook will be helpful in this respect.

The data for each example are read from a separate input file in which the data (often including labels) are structured exactly as presented in *App. Stat.*, thereby demonstrating some of the flexibility of Genstat. The input procedure is explained on pp. 4–5; beginners wishing to keep to more simple input formats should easily be able to do so by following some of the simpler examples.

Of the examples analysed, D and E, although containing interesting discussion in *App. Stat.*, are computationally simple and straightforward requiring only plotting and linear regression; as such, they provide an easy introduction to Genstat. Examples G and P involve straightforward multiple regression. Example V requires only READ, CALCULATE and PRINT commands. Other more complex problems involve the fitting of logistic models (Examples H and X), loglinear models (Example W), gamma distributions (Example T) and nonlinear models (Example U). Residual, or restricted, maximum likelihood estimation (REML) is used for models involving components of variance in Examples I, Q, and S).

For each example, the Genstat program has been run in batch mode and under the heading PROGRAM the resulting output file is shown. This lists the input, output and comments (which are enclosed in quotes and are ignored by the program); see OUTPUT (p. 5) for the construction of a transcript file for interactive working.

All programs listed in this handbook have been run on Genstat 5, Release 2.

An introduction to the Genstat statistical system

Genstat is fully described in the Genstat 5 Reference Manual (OUP 1987); here only a concise outline is given, sufficient to help the user follow the analyses in Examples A, ..., X. Readers wishing for greater detail should consult *Genstat 5: An Introduction* (Lane, Galwey and Alvey, OUP 1987) and *Genstat 5: A Second Course* (Digby, Galwey and Lane, OUP 1989).

THE GENSTAT LANGUAGE

A simple example of a complete job is

```
JOB       'Areas of circles'
VARIATE   RADIUS; !(1...5)
CALCULATE AREA=3.14159*RADIUS**2
PRINT     RADIUS,AREA
STOP
```

RADIUS is declared as a variate and assigned values 1 to 5 (the abbreviation p,q ... r defines an arithmetical progression p, p+d, p+2d ... r where d=q−p; if |d|=1 the second term may be omitted). The CALCULATE command implicitly declares AREA as a variate of the same length as RADIUS and computes the area for each value of RADIUS. PRINT then displays the values of the two variates in columns, with 12 spaces allowed for each (the number of decimal places used is such that the mean absolute value would be printed with four significant figures).

The names of the commands (JOB, VARIATE, ...) are system words which may appear in either upper or lower case. The names RADIUS and AREA are supplied by the user; either case may be used but RADIUS, Radius and radius will normally be treated as different names. Note, for clarity of presentation, we have used upper case throughout this handbook for all Genstat commands and user names.

They thus stand out clearly from comments, enclosed in quotes (". . ."), which are intended as a help to the reader and are interspersed in the programs for the Examples.

Each Genstat statement has a defined action and standard analyses may be specified easily. The default action can be varied by supplying options and parameter lists.

The list RADIUS, AREA is a parameter of the PRINT command; other parameter lists, which may be named, can be used to override the default format. For example,

```
PRINT    RADIUS,AREA; FIELDWIDTH=9; DECIMALS=0,2
```

will display the values of RADIUS as integers and those of AREA to two decimal places in fields of width 9. Parameter lists are used in parallel with the first list (which must be the longest); shorter lists are recycled as often as necessary.

Option settings are given in square brackets after the command name and apply to the whole statement. Thus,

```
PRINT    [ORIENT=ACROSS]  RADIUS,AREA; FIELDWIDTH=9; DECIMALS=0,2
```

will print the values in rows instead of columns.

Genstat may be run interactively or in batch mode. When working interactively the user will see the input commands as they are typed, followed by any output generated by a command; in batch mode, with output sent to file, input lines are numbered and the output from this job (with the two forms of the PRINT command) will be as shown below.

```
1  JOB        'Areas of circles'
2  VARIATE    RADIUS; !(1...5)
3  CALCULATE  AREA=3.14159*RADIUS**2
4  PRINT      RADIUS,AREA

   RADIUS       AREA
    1.000       3.14
    2.000      12.57
    3.000      28.27
    4.000      50.27
    5.000      78.54

5  PRINT      [ORIENT=ACROSS]  RADIUS,AREA; FIELDWIDTH=9; DECIMALS=0,2

   RADIUS      1        2        3        4        5
     AREA    3.14    12.57    28.27    50.27    78.54

6  STOP
```

A line number is preceded by a negative sign (−) when within strings (enclosed by primes (')) or within comments (enclosed by quotes (")), and this is the form in which the programs for Examples A, ..., X are shown.

Statements may be extended over more than one line by using a continuation symbol (\). Several statements may be given on the same line, separated by : (colon) when a different command follows or & (ampersand) which repeats the previous command and its option settings. Spaces may be used between items. System words need not be given in full; this is particularly useful when working interactively. Instances of these will be found in the Examples in this handbook.

DATA STRUCTURES

Several types of one-dimensional structures are recognised:

1. variates, of real numbers;
2. factors (qualitative or categorical variates);
3. texts, of strings of characters; and
4. pointers, whose values are references to other structures.

The statement

POINTER   P; !P(X,Y,Z)

sets up such a structure; P[2] points to Y and P[] refers to all (current) members of the pointer P and is equivalent here to the list X, Y, Z.

A UNITS command is often used to set a default length for one-dimensional structures.

UNITS   [35]

sets the default length to 35.

Several forms of matrix (rectangular, symmetric and diagonal) and multiway tables with up to nine dimensions are accepted.

Strings, unless they begin with a letter and consist only of letters and digits, must be enclosed in primes ('); see the JOB name above. When a text is printed, each of its strings appear on a separate line.

The concept of a factor is important, particularly for the analysis of designed experiments and survey data. The command

FACTOR   [NLEVELS=4] PHOSPHATE

declares PHOSPHATE as a factor with four (formal) levels 1, 2, 3, and 4. These are the set of legal values for the factor. The number of levels must not be confused with the number of values held in the structure (in an experiment this will be the number of plots, in a survey, the number of units).

There are two other ways of referring to the levels of a factor: by labels and by actual (numerical) levels.

```
FACTOR    [LABELS=!T(N0,N1,N2); LEVELS=!(0.0,12.5,25.0)] NITROGEN
```

sets up NITROGEN as a factor with three levels, labels N0, N1 and N2 and actual levels 0.0, 12.5 and 25.0. The data may be punched as formal or actual levels or labels; labels, if specified, are used in output. In calculations the actual levels are used if they have been defined, if not formal levels will be used.

Tables are defined by sets of factors, the numbers of levels determining the size of each dimension. Space for marginal values may be reserved.

ASSIGNING VALUES TO STRUCTURES

Values may be assigned to structures at declaration, by input (see READ, below), or by commands e.g. CALCULATE. If a set of values is needed only for the current command an 'unnamed structure' may be used, with the values enclosed in brackets, preceded by ! as in the VARIATE statement in the first example. In the FACTOR statement above, the values assigned to the LABELS option are strings; the letter T after the ! tells the program that strings are expected.

The symbol #, used before an identifier, causes the identifier to be replaced by its values. For example, if Y and Z already have numerical values,

```
VARIATE  X;  !(#Y,#Z)
```

sets up a variate X containing the values of Y followed by those of Z.

Subsets of values of a structure may be referenced: RADIUS$[3] refers to the third element of RADIUS; if J has values (1,4), RADIUS$[J] is a variate of length 2 consisting of the first and fourth elements of RADIUS.

INPUT

Data values are input using the READ command. This has a number of options and parameters which allow data to be presented in any manner, as characters in fixed or free-format or unformatted, from files written by Genstat or other Fortran programs. With one exception, the data for the analyses discussed in this book are in a form that is easily read (and checked), with spaces separating the values. This allows the simplest form of READ to be used, the only complication arising when factors are read. The default assumption is that actual levels will be found; to accept labels, a parameter FREPRESENTATION has to be set, e.g.

```
READ      NITROGEN; FREPRESENTATION=LABELS
```

The input of fixed-format data is illustrated in Example P.

For each of the examples A, . . ., X the input data is in separate files called XA, . . ., XX, respectively. These are read from a secondary input channel accessed by

```
OPEN      'X-'; 2
```

where – is replaced by the appropriate letter A, . . ., X. The form of the data is shown as a comment following the OPEN statement. Normally the data are presented as they appear in *App. Stat.* but in some cases (Examples K, L, W, and X) where the values of a factor follow a regular pattern it is more sensible to use a GENERATE command to set up the factor values. Note that an end-of-data marker (:) is required after each set of structure values punched in parallel (unit by unit) or after the values of each structure when punched serially.

OUTPUT

The PRINT command has many options and parameters which allow the user to control the appearance of the output, many examples of which appear in this handbook.

The COPY command may be used to generate a transcript file of a job when working interactively by using the following code.

```
OPEN      'filename'; CHANNEL=3; FILETYPE=OUTPUT
COPY      [PRINT=STATEMENTS]  3
```

High-quality graphics are produced using the DGRAPH command. Before this command is used, a graphics output file must be opened, a graphics device nominated, and the graphics environment established by using AXES, FRAME, PENS and WINDOWS commands. All these depend on the facilities available at a particular site, and the versions used in this handbook (in Examples A, B, C, H, K and U) may need to be modified. Line-printer graphs are produced using the GRAPH command in Examples A, B, D, E, F, G and T.

DATA MANIPULATION

After values have been assigned to structures, either directly at declaration or from input, new values can be derived by the CALCULATE command. This will evaluate expressions using the familiar mathematical symbols and a wide range of mathematical and statistical functions. Thus in

```
CALCULATE A = LOG10(B+0.1)
```

A and B may be of any type; they must of course be compatible: if variates, of the same length, if matrices, of the same shape. A single structure may be replaced by a list of structures: the first list must be the longest and shorter lists are recycled.

```
CALCULATE A1,A2,A3 = LOG10(B1,B2,B3+0.1,0.5)
```

will add 0.1 to B1 and B3, and 0.5 to B2, before taking logarithms.

The familiar matrix operations (multiply, transpose, invert etc.) are provided as functions. A powerful table calculus is included: tables of different shapes (i.e. classified by different sets of factors) may be combined, marginal totals over factors not common to the classification sets being used when necessary.

STATISTICAL ANALYSES

When the data structures have been established, and any transformations required have been calculated, many standard statistical techniques are available.

```
MODEL    Y
FIT      X1,X2
```

will fit the linear regression of Y on X1 and X2, under the usual assumption of normal errors. Generalised linear models with various error distributions (Examples H, M, W and X) and nonlinear models may be fitted, and functions minimised (Example T). The ADD or DROP commands will add or drop variables from the model (Examples F, G, J, L and M). The STEP command can be used for backward elimination or forward selection of variables (Example P).

To analyse data from a balanced experiment, two preliminary commands are needed: BLOCK to define the randomisation and TREATMENT to specify the treatment model to be fitted (Examples I to N, Q, R and S). These are expressed as formulae involving factors; the error strata and confounding are worked out from these formulae and are not given explicitly (Example M).

The standard techniques for multivariate analysis are available, and time series may be analysed using Box–Jenkins models. However no examples illustrating multivariate or time series analysis are given here.

OTHER FEATURES OF THE LANGUAGE

Genstat is a structured programming language: FOR–ENDFOR loops, and IF–ENDIF and CASE blocks are supported. FOR loops are used in Examples C, N, P, T and W; IF blocks in Example S.

Genstat language routines may be stored in procedure libraries. When a procedure has been defined, it is invoked by using the procedure name, options and parameters as if it were a standard command. A procedure library containing many useful extensions to the standard facilities is supplied with the program: ORTHPOL (which generates orthogonal polynomials) is used in Examples F, J, L and M, and RCHECK (to check the results of a regression analysis) is used in Example G.

POSTSCRIPT

Genstat is distributed by the Numerical Algorithms Group and enquiries should be addressed to:

> NAG Ltd.
> Wilkinson House
> Jordan Hill Road
> Oxford OX2 8DR

Examples

Example A

Admissions to intensive care unit

DESCRIPTION OF DATA

Table A.1 gives arrival times of patients at an intensive care unit. The data were collected by Dr. A. Barr, Oxford Regional Hospital Board. Interest lies in any systematic variations in arrival rate, especially any that might be relevant in planning future administration.

THE ANALYSIS

It would be simple to calculate by hand the frequency distributions for time-of-day, day-of-week, and calendar-month (Tables A.2, A.3 and A.4, *App. Stat.*) but calculation of the time intervals between admissions, particularly if Table A.5, *App. Stat.* were extended, is less tedious done by computer. While reading in the data to construct one frequency table, it is sensible to construct others at the same time.

The data are input in the form shown in Table A.1, with months contracting to three letters and losing the full stop.

The three-way table for time × day × month is constructed first, from which the marginal frequency distributions for time-of-day, day-of-week and calendar-month are obtained.

The chi-squared goodness-of-fit statistic, assuming a constant arrival rate on each day-of-week (Table A.3 *App. Stat.*) is computed. Table A.4 *App. Stat.*, showing the long-term variation over months, is also reproduced.

Time-of-day variation is illustrated using high-quality graphics; see Fig. A.1 after the Genstat program.

Table A.1 Arrival times of patients at intensive care unit (to be read down the columns.)

1963			1963			1963			1963		
M	4 Feb.	11.00 hr	Th	28 Mar.	12.00 hr	W	22 May	22.00 hr	W	26 June	6.30 hr
		17.00			12.00			10.15			17.30
F	8 Feb.	23.15	S	30 Mar.	18.00	Th	23 May	12.30	Th	27 June	20.45
M	11 Feb.	10.00	T	2 Apr.	22.00	F	24 May	18.15	S	29 June	22.00
S	16 Feb.	12.00	S	6 Apr.	22.00	S	25 May	21.05	Su	30 June	20.15
M	18 Feb.	8.45	T	9 Apr.	22.05			21.00	T	2 July	21.00
		16.00	W	10 Apr.	12.45	T	28 May	0.30			17.30
W	20 Feb.	10.00	Th	11 Apr.	19.30			1.45	M	8 July	19.50
		15.30	M	15 Apr.	18.45	Th	30 May	12.20	T	9 July	2.00
Th	21 Feb.	20.20	T	16 Apr.	16.15			14.45	W	10 July	1.45
M	25 Feb.	4.00			16.00	S	3 June	22.30	F	12 July	3.40
		12.00	T	23 Apr.	20.30	M	3 June	12.30	S	13 July	4.15
Th	28 Feb.	2.20	Su	28 Apr.	23.40	W	5 June	13.15			23.55
F	1 Mar.	12.00	M	29 Apr.	20.20			17.30			3.15
Su	3 Mar.	5.30	S	4 May	18.45	M	10 June	11.20	S	20 July	19.00
Th	7 Mar.	7.30	M	6 May	16.30			17.30	Su	21 July	21.45
		12.00	T	7 May	22.00	W	12 June	23.00	T	23 July	21.30
S	9 Mar.	16.00			8.45	Th	13 June	10.55	W	24 July	0.45
F	15 Mar.	16.00	S	11 May	19.15			13.30			2.30
S	16 Mar.	1.30	M	13 May	15.30	Su	16 June	11.00	S	27 July	15.30
Su	17 Mar.	11.05	T	14 May	12.00	T	18 June	18.30	M	29 July	21.00
W	20 Mar.	16.00	Th	16 May	18.15	F	21 June	11.05			8.45
F	22 Mar.	19.00	S	18 May	14.00	S	22 June	4.00	Th	1 Aug.	14.30
Su	24 Mar.	17.45	Su	19 May	13.00			7.30	F	2 Aug.	17.00
		20.20	M	20 May	23.00	M	24 June	20.00	S	3 Aug.	3.30
		21.00			19.15	Tu	25 June	21.30	W	7 Aug.	15.45

Table A.1 (*cont.*)

1963			1963			1963			1963		
Su	11 Aug.	17.30 hr	S	28 Sept.	17.30 hr	S	9 Nov.	13.45 hr	Th	5 Dec.	10.05 hr
T	13 Aug.	14.00	T	1 Oct.	12.30	M	11 Nov.	19.30			20.00
		2.00	W	2 Oct.	17.30	T	12 Nov.	0.15	S	7 Dec.	13.35
		11.30	Th	3 Oct.	14.30	F	15 Nov.	7.45			16.45
		17.30			16.00			15.20	Su	8 Dec.	2.15
M	19 Aug.	17.10	Su	6 Oct.	14.10			18.40	M	9 Dec.	20.30
W	21 Aug.	21.20	T	8 Oct.	14.00	S	16 Nov.	19.50	W	11 Dec.	14.00
S	24 Aug.	3.00	S	12 Oct.	15.30	Su	17 Nov.	23.55	Th	12 Dec.	21.15
S	31 Aug.	13.30	Su	13 Oct.	4.30	M	18 Nov.	1.45	F	13 Dec.	18.45
M	2 Sept.	23.00	S	19 Oct.	11.50	T	19 Nov.	10.50	S	14 Dec.	14.05
Th	5 Sept.	20.10	Su	20 Oct.	11.55	F	22 Nov.	7.50			14.15
S	7 Sept.	23.15			15.20	S	23 Nov.	15.30	Su	15 Dec.	1.15
Su	8 Sept.	20.00	T	22 Oct.	15.40			18.00	M	16 Dec.	1.45
T	10 Sept.	16.00	W	23 Oct.	11.15	Su	24 Nov.	23.05	T	17 Dec.	18.00
W	11 Sept.	18.30	S	26 Oct.	2.15	T	26 Nov.	19.30	F	20 Dec.	14.15
F	13 Sept.	21.00	W	30 Oct.	11.15	W	27 Nov.	19.00			15.15
Su	15 Sept.	21.10	Th	31 Oct.	21.30	F	29 Nov.	16.10	S	21 Dec.	16.15
M	16 Sept.	17.00	F	1 Nov.	3.00	S	30 Nov.	10.00	Su	22 Dec.	10.20
W	18 Sept.	13.25			0.40			2.30	M	23 Dec.	13.35
S	21 Sept.	15.05	M	4 Nov.	10.00	Su	1 Dec.	22.00			17.15
M	23 Sept.	14.10			9.45	M	2 Dec.	21.50	T	24 Dec.	19.50
T	24 Sept.	19.15	T	5 Nov.	23.45	T	3 Dec.	19.10			22.45
		14.05	W	6 Nov.	10.00			11.45	W	25 Dec.	7.25
		22.40	Th	7 Nov.	7.50			15.45			17.00
F	27 Sept.	9.30	F	8 Nov.	13.30			16.30	S	28 Dec.	12.30
					12.30			18.30	T	31 Dec.	23.15

Table A.1 (*cont.*)

1964			1964			1964			1964		
Th	2 Jan.	10.30 hr	W	15 Jan.	18.35 hr	Th	6 Feb.	23.10 hr	Su	23 Feb.	2.30 hr
F	3 Jan.	13.45	F	17 Jan.	13.30	F	7 Feb.	19.15	M	24 Feb.	12.55
Su	5 Jan.	2.30	Su	19 Jan.	16.40	T	11 Feb.	11.00	T	25 Feb.	20.20
M	6 Jan.	12.00	M	20 Jan.	18.00			0.15	W	26 Feb.	10.30
T	7 Jan.	15.45	T	21 Jan.	20.00	W	12 Feb.	14.40	M	2 Mar.	15.50
		17.00	F	24 Jan.	11.15	M	17 Feb.	15.45	W	4 Mar.	17.30
		17.00	S	25 Jan.	16.40	T	18 Feb.	12.45	F	6 Mar.	20.00
F	10 Jan.	1.30	W	29 Jan.	13.55			17.00	T	10 Mar.	2.00
		20.15	Th	30 Jan.	21.00			18.00	W	11 Mar.	1.45
S	11 Jan.	12.30	F	31 Jan.	7.45	W	19 Feb.	21.45	W	18 Mar.	1.45
Su	12 Jan.	15.40	W	5 Feb.	22.30	Th	20 Feb.	16.00			2.05
T	14 Jan.	3.30			16.40			12.00			

PROGRAM

```
 2  JOB       'Example A'
 3
 4  OPEN      'XA'; 2
 5
 6  "All the data for 1963 are punched first, followed by the 1964 figures,
-7  beginning with
-8  M   4 Feb 11.00    T   2 Apr 22.00    S   1 Jun 12.20    S  13 Jul 23.55
-9  and ending
-10 T  14 Jan  3.30   W   5 Feb 16.40    Th 20 Feb 12.00:
-11 The information for year is given directly, by declaring a factor
-12 with appropriate values:"
13
14  UNITS     [254]
15  FACTOR    [LEVELS=2]  YEAR; !(20/(1),47(2))
16
17  "Declare symbols used for Days and Months in data."
18
19  TEXT      NDAY; !T(M,T,W,Th,F,S,Su)
20  TEXT      NMON; !T(Jan,Feb,Mar,Apr,May,Jun,Jul,Aug,Sep,Oct,Nov,Dec)
21  FACTOR    [LABELS=NDAY]  DAY
22  &         [LABELS=NMON]  MONTH
23  FACTOR    [LEVELS=31]  DATE
24  READ      [CHANNEL=2]  DAY,DATE,MONTH,VTIME; FREP=Labels,*
```

Identifier	Minimum	Mean	Maximum	Values	Missing
VTIME	0.15	14.13	23.55	254	0

```
25
26  "Define output labels for Day and Time and group time into two-hourly intervals"
27
28  TEXT      NDAY; !T(Mon,Tue,Wed,Thu,Fri,Sat,Sun)
29  TEXT      NTIM; !T(' 0.00-',' 2.00-',' 4.00-',' 6.00-',' 8.00-','10.00-',     \
30                    '12.00-','14.00-','16.00-','18.00-','20.00-','22.00-')
31  SORT      [INDEX=VTIME; GROUPS=TIME; LIMITS=!(1.99,3.99...21.99)]
32  FACTOR    [MODIFY=Y; LABELS=NTIM] TIME
33
34  "Create a factor PERIOD to represent the 14 months of the survey."
35
36  TEXT      NPER; !T('Feb 63','Mar 63','Apr 63','May 63','Jun 63',              \
37                    'Jul 63','Aug 63','Sep 63','Oct 63','Nov 63',               \
38                    'Dec 63','Jan 64','Feb 64','Mar 64')
39  FACTOR    [LABELS=NPER] PERIOD
40  CALCULATE PERIOD = MONTH + 12*YEAR - 13
41
42  "Form three-way table of arrivals by
-43  period (month), day (of week) and time (of day)."
44
45  TABULATE  [CLASS=PERIOD,DAY,TIME; COUNT=ARRIVALS]
46
47  "This table can be printed if required by
-48 PRINT     ARRIVALS; FIELD=8; DECIMALS=0
-49
-50 Extract the one-way table of arrivals by time-of-day and compute rates:"
51
52  TABLE     [CLASS=TIME] ATD,RTD
53  CALCULATE ATD = ARRIVALS
54  &         RTD = ATD/(409/12)
55
56  " Time-of-day variation:
-57  ATD is the number of arrivals, RTD the rate per day."
58
59  PRINT     ATD,RTD; FIELD=12; DECIMALS=0,3              "Table A.2 App. Stat"
```

TIME	ATD	RTD
0.00-	14	0.411
2.00-	17	0.499
4.00-	5	0.147
6.00-	8	0.235
8.00-	5	0.147

```
        10.00-        25       0.733
        12.00-        31       0.910
        14.00-        30       0.880
        16.00-        36       1.056
        18.00-        29       0.851
        20.00-        31       0.910
        22.00-        23       0.675
```

```
60    "Similarly, for day-of-week. Set up a variate containing the number of times
-61   each day occurs and calculate chi-squared:"
62
63    TABLE      [CLASS=DAY] ADW,FDW,RDW
64      &        NDW; !(3(59),4(58))
65    CALCULATE ADW = ARRIVALS
66      &        RDW = ADW/NDW
67      &        FDW = SUM(ADW)*NDW/SUM(NDW)
68      &        CS = SUM( (FDW-ADW)**2/FDW )
69      &        DF = NVAL(NDW)-1
70
71    " Day-of-week variation:
-72     NDW is the number of weeks, ADW the number of arrivals;
-73     FDW is the fitted frequencey and RDW the rate per day."
74
75    PRINT      NDW,ADW,FDW,RDW; FIELD=12; DECIMALS=0,0,2,3      "Table A.3 App. Stat."
```

	NDW	ADW	FDW	RDW
DAY				
Mon	59	37	36.64	0.627
Tue	59	53	36.64	0.898
Wed	59	35	36.64	0.593
Thu	58	27	36.02	0.466
Fri	58	30	36.02	0.517
Sat	58	44	36.02	0.759
Sun	58	28	36.02	0.483

```
76    TEXT       T
77    PRINT      [CHANNEL=T; IPRINT=*; SQUASH=Y] '   Chisquare =',CS,'  (df:',DF,')';\
78               FIELD=12,5,9,2,1; DECIMALS=0,2,0,0,0
79    PRINT      [IPRINT=*] T

      Chisquare = 14.20     (df: 6 )

80    "To ignore a given day, e.g. Tuesday, EQUATE may be used to set the required
-81   values of NDW and ADW to variates of length 6.  Details are not shown here.
-82
-83   Repeat the exercise for long-term variation:"
84
85    TABLE      [CLASS=PERIOD] ALT,FLT,RLT
86      &        NLT; !(25,31,30,31,30,31,31,30,31,30,31,31,29,18)
87    CALCULATE ALT = ARRIVALS
88      &        RLT = ALT/NLT
89      &        FLT = SUM(ALT)*NLT/SUM(NLT)
90      &        CS = SUM( (FLT-ALT)**2/FLT )
91      &        DF = NVAL(NLT)-1
92
93    " Long-term variation:
-94     NLT is the number of days, ALT the number of arrivals;
-95     FLT is the fitted frequency, and RLT is the rate per day."
96
97    PRINT      NLT,ALT,FLT,RLT; FIELD=12; DEC=0,0,2,3      "Table A.4 App. Stat."
```

	NLT	ALT	FLT	RLT
PERIOD				
Feb 63	25	13	15.53	0.520
Mar 63	31	16	19.25	0.516
Apr 63	30	12	18.63	0.400
May 63	31	18	19.25	0.581
Jun 63	30	23	18.63	0.767
Jul 63	31	16	19.25	0.516
Aug 63	31	15	19.25	0.484

```
        Sep 63        30        17        18.63        0.567
        Oct 63        31        17        19.25        0.548
        Nov 63        30        28        18.63        0.933
        Dec 63        31        32        19.25        1.032
        Jan 64        31        23        19.25        0.742
        Feb 64        29        17        18.01        0.586
        Mar 64        18         7        11.18        0.389

 98  PRINT      [CHANNEL=T; IPRINT=*; SQUASH=Y] '    Chisquare =',CS,'  (df:',DF,')';\
 99             FIELD=12,5,9,2,1; DECIMALS=0,2,0,0,0
100  PRINT      [IPRINT=*] T

     Chisquare = 21.82      (df: 13 )

101  "A line printer graph showing time-of-day variation can be obtained by setting
-102  up suitable variates."
103
104  VARIATE    PAR; !(10(#RTD))
105  &          TOD; !(0.00,0.20...23.80)
106
107  "and then using the GRAPH command
-108  GRAPH      PAR; TOD; SYMBOL=':'
-109
-110  Here instead we illustrate the use of high-quality Graphics:"
111
112  OPEN       'A.PLT'; 6; Graphics
113  DEVICE     6
114  AXES       WINDOW=1; YTITLE='Arrival rate per day'; XTITLE='Time of day';    \
115             YLOWER=0.0; YUPPER=1.5; XLOWER=0; XUPPER=25; XINTEGER=Y;           \
116             XMARKS=!(0,2...24)
117  PEN        1; SYMBOL = '-'
118  DGRAPH     [TITLE=' Patient arrival rate versus time of day'; KEYWINDOW=0]    \
119             PAR; TOD; PEN=1
120  "To investigate the intervals between arrivals, the data must be sorted
-121  into the correct order. Construct a variate TD containing the total number
-122  of days from February 1st 1963 to the beginning of each period. By setting
-123  this variate as the levels vector of the factor PERIOD, and converting the
-124  representation of time in VTIME (hours.min) to true decimal form, the times
-125  of arrivals can be constructed as hours from the origin."
126
127  VARIATE    TD; !(0,28,31,30,31,30,31,31,30,31,30,31,31,29)
128  CALCULATE  TD = CUMULATE(TD)
129  FACTOR     [MODIFY=Y; LEVELS=TD] PERIOD
130  CALCULATE  AT = (DATE+PERIOD)*24 + INTEGER(VTIME)+(VTIME-INTEGER(VTIME))*100/60
131
132  "The arrival times can now be sorted, differenced, grouped and tabulated:"
133
134  SORT       AT
135  CALCULATE  AT = DIFFERENCE(AT)
136  SORT       [INDEX=AT; GROUPS=DIFFS; LIMITS=!(1.9999,3.9999,5.9999)]
137  TEXT       NDIF; !T('0 hr-  ','2 hr-  ','4 hr-  ','6 hr+ ')
138  FACTOR     [MODIFY=Y; LABELS=NDIF] DIFFS
139  TABULATE   [CLASS=DIFFS; COUNT=INTERVAL]
140  PRINT      INTERVAL; FIELD=5; DECIMALS=0                    "Table A.5 App. Stat."

               INTERVAL
        DIFFS
        0 hr-      20
        2 hr-      13
        4 hr-       6
        6 hr+     214

Unknown    INTERVAL    1

141  "Note that early editions of App. Stat. contain an error in Table A.5."
142
143  STOP

******** End of Example A.  Maximum of 16448 data units used at line 130 (33542 left)
```

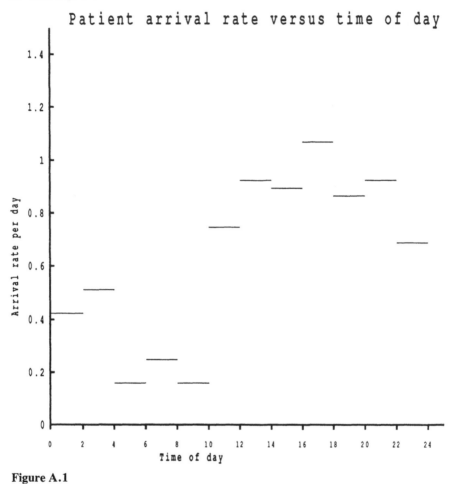

Figure A.1

SUGGESTED FURTHER WORK

For the time intervals between successive admissions:

1. extend the frequency distribution by grouping into narrower intervals and splitting the final group;
2. fit an appropriate theoretical distribution and examine its adequacy.

Example B

Intervals between adjacent births

DESCRIPTION OF DATA

The data in Table B.1 were obtained by Greenberg and White (1963) from the records of a Genealogical Society in Salt Lake City. The distribution of intervals between successive births in a particular serial position is approximately log normal and hence geometric means are given. For example, the entry 39.9 in the bottom right-hand corner means that for families of 6 children, in which the fifth and sixth children are both girls, the geometric mean interval between the births of these two children is 39.9 months.

Table B.1 Mean intervals in months between adjacent births by family size and sequence of sex at specified adjacent births

Family size	Births	Sequence of sex			
		MM	MF	FM	FF
2	1–2	39.8	39.5	39.4	39.3
3	1–2	31.0	31.5	31.4	31.1
3	2–3	42.8	43.7	43.3	43.4
4	1–2	28.4	28.1	27.5	27.8
4	2–3	34.2	34.4	34.3	35.0
4	3–4	43.1	44.3	43.3	42.8
5	1–2	25.3	25.6	25.6	25.5
5	2–3	30.3	30.1	29.9	30.0
5	3–4	33.7	34.0	33.7	34.7
5	4–5	41.6	42.1	41.9	41.3
6	1–2	24.2	24.4	24.0	24.5
6	2–3	27.6	27.7	27.5	27.6
6	3–4	29.8	30.2	30.3	30.8
6	4–5	34.2	34.2	34.1	33.4
6	5–6	40.3	41.0	40.6	39.9

THE ANALYSIS

The analysis here is at a very descriptive level and the amount of data so small that it is doubtful whether a computer run really is justified. However, the example provides a useful illustration of the following:

1. reading text containing symbols other than numbers or letters;
2. logarithmic and exponential transformation;
3. calculation of means and ranges;
4. line-printer and high-quality graphics plots, as in Fig. B.1.

PROGRAM

```
 2  JOB        'Example B'
 3
 4  OPEN       'XB'; 2
 5
 6  "Data:
-7  2 '1-2'   39.8 39.5 39.4 39.3
-8  ...
-9  6 '5-6'   40.3 41.0 40.6 39.9 :
-10
-11 Declare symbols used for family size and birth intervals in input data."
12
13  UNITS      [15]
14  FACTOR     [LEVELS=!(2...6)] SIZE
15  FACTOR     [LABELS=!T('1-2','2-3','3-4','4-5','5-6')] BIRTHS
16  POINTER    [NVAL=4] DATA; !P(MM,MF,FM,FF)
17  READ       [CHANNEL=2] SIZE,BIRTHS,DATA[]; FREP=Levels,Labels,4(*)
```

Identifier	Minimum	Mean	Maximum	Values	Missing
MM	24.20	33.75	43.10	15	0
MF	24.40	34.05	44.30	15	0
FM	24.00	33.79	43.30	15	0
FF	24.50	33.81	43.40	15	0

```
18
19  "In order to examine variation across sequence of sex (Fig B.1, App. Stat.)
-20 we transfer the data values to a 4*15 matrix, transpose it, and extract the
-21 rows into 15 variates of length 4."
22
23  POINTER    [NVAL=15] LINE
24  VARIATE    [NVAL=4] LINE[]
25  MATRIX     [ROW=4; COLUMN=15] M1
26  &          [ROW=15; COLUMN=4] M2
27  EQUATE     DATA; M1
28  CALCULATE  M2 = TRANSPOSE(M1)
29  EQUATE     M2; LINE
30
31  "To label each line in a high-quality graph with a different symbol, the
-32 symbols have to be assigned to different pens. Three DGRAPH statements are
-33 used, redefining pens and using the option SCREEN=KEEP of DGRAPH to accumulate
-34 one picture."
35
36  OPEN       'B.PLT'; 6; Graphics
37  DEVICE     6
38  AXES       WINDOW=1; XTITLE='Sequence of sex'; YTITLE='Months';          \
39             YLOWER=21; YUPPER=50; XLOWER=0; XUPPER=5; YINTEGER=Y; XINTEGER=Y;  \
40             XMARKS=!(1...4); XLABELS=!T(MM,MF,FM,FF)
41
42  PEN        2,3,4,6,7; LINESTYLE=2; METHOD=LINE; SYMBOLS='a','c','i','b','d'
43  DGRAPH     [KEYWINDOW=0] LINE[1,3,9,2,4]; !(1,2,3,4); PEN=2,3,4,6,7
44  PEN        2,3,4,7,8; LINESTYLE=2; METHOD=LINE; SYMBOLS='e','g','j','h','k'
45  DGRAPH     [SCREEN=KEEP; KEYWINDOW=0]                                     \
46             LINE[5,7,10,8,11]; !(1,2,3,4); PEN=2,3,4,7,8
```

```
47  PEN        2,3,4,6,7; LINESTYLE=2; METHOD=LINE; SYMBOLS='f','n','m','l','o'
48  DGRAPH     [SCREEN=KEEP; KEYWINDOW=0]                                        \
49             LINE[6,14,13,12,15]; !(1,2,3,4); PEN=2,3,4,6,7
50
51  POINTER    [NVAL=4]  LOGDATA
52  CALCULATE  LOGDATA[] = LOG(DATA[])
53     &       LOGMONTH  = VMEAN(LOGDATA)
54     &       MONTHS    = EXP(LOGMONTH)
55     &       MEANR     = MEAN(VMAX(LOGDATA)-VMIN(LOGDATA))
56
57  "Average values of logmonths, and corresponding antilogs."
58
59  PRINT      SIZE,BIRTHS,LOGMONTH,MONTHS; FIELD=8,8,10,10; DECIMALS=0,0,3,1     \
60     &       MEANR; FIELD=13; DEC=4                        "Table B.2 App. Stat."
```

SIZE	BIRTHS	LOGMONTH	MONTHS
2	1-2	3.676	39.5
3	1-2	3.442	31.2
3	2-3	3.768	43.3
4	1-2	3.330	27.9
4	2-3	3.540	34.5
4	3-4	3.770	43.4
5	1-2	3.239	25.5
5	2-3	3.404	30.1
5	3-4	3.527	34.0
5	4-5	3.731	41.7
6	1-2	3.189	24.3
6	2-3	3.318	27.6
6	3-4	3.410	30.3
6	4-5	3.526	34.0
6	5-6	3.700	40.4

```
        MEANR
       0.0216
```

```
61  CALCULATE ORDER = BIRTHS - SIZE + 1
62  GRAPH      [YTITLE='Log months'; XTITLE='Order in family'; XLOWER=-5; XUPPER=1; \
63             NROW=22; NCOL=64; XINTEGER=Y]  LOGMONTH; ORDER; SYMBOLS = BIRTHS
```

```
       -+---------+---------+---------+---------+---------+---------+----
        I                                                              I
        I                                                              I
        I                                                              I
  3.75  I                                                    :         I
        I                                                    5-        I
        I                                                    1-        I
L       I                                                              I
o       I                                                              I
g       I                                            :                 I
  3.50  I                                                              I
m       I                                          1-                  I
o       I                                     :                        I
n       I                                                              I
t       I                    2-          1-                            I
h       I                                                              I
s 3.25  I                    1-                                        I
        I          1-                                                  I
        I                                                              I
        I                                                              I
        I                                                              I
        I                                                              I
  3.00  I                                                              I
       -+---------+---------+---------+---------+---------+---------+----
     -5.000    -4.048    -3.095    -2.143    -1.190    -0.238    0.714
```

Order in family

```
64  "0: final interval, -1: previous interval, -2: 2 before final, ...."
```

```
-65    (In the graph, a symbol ':' indicates coincident points)."
 66
 67    RESTRICT  MONTHS,LOGMONTH; BIRTHS.EQ.1
 68
 69    "Intervals between births 1-2:"
 70
 71    PRINT     SIZE,LOGMONTH,MONTHS;                                           \
 72              FIELD=8,10,10; DECIMALS=0,3,1              "Table B.3 App. Stat."
```

```
     SIZE  LOGMONTH    MONTHS
        2    3.676       39.5
        3    3.442       31.2
        4    3.330       27.9
        5    3.239       25.5
        6    3.189       24.3
```

```
 73    FACTOR    [LABELS=!T('Final interval','1 before final','2 before final',  \
 74              '3 before final','4 before final')] INTERVAL
 75    CALCULATE INTERVAL = ABS(ORDER)+1
 76    TABLE     [CLASS=INTERVAL] TI,LOGTI
 77    RESTRICT  MONTHS,LOGMONTH; (SIZE.GT.2).AND.(BIRTHS.GT.1)
 78    TABULATE  LOGMONTH; MEANS=LOGTI
 79    CALCULATE TI = EXP(LOGTI)
 80
 81    "Intervals between successive births
-82    (families size 3 or more, intervals 1-2 excluded):"
 83
 84    PRINT     LOGTI,TI; FIELD=10; DECIMALS=3,1         "Table B.3 App. Stat., Cont."
```

```
                    LOGTI        TI
          INTERVAL
Final interval      3.742       42.2
1 before final      3.531       34.2
2 before final      3.407       30.2
3 before final      3.318       27.6
4 before final        *           *
```

```
 85    STOP
```

```
******** End of Example B.  Maximum of 13934 data units used at line 75 (36056 left)
```

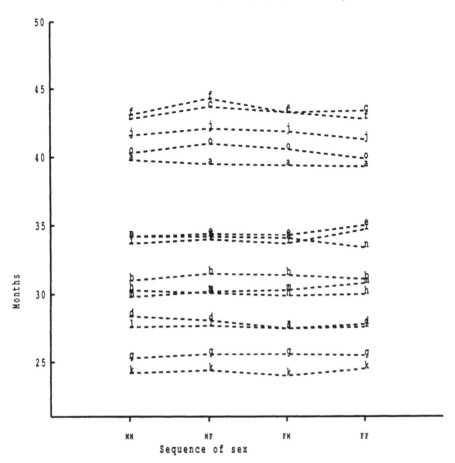

Figure B.1

Example C

Statistical aspects of literary style

DESCRIPTION OF DATA

As part of an investigation of the authorship of works attributed to St. Paul, Morton (1965) found the numbers of sentences having zero, one, two, ... occurrences of 'kai' ($\equiv$ and) in some of the Pauline works. Table C.1 gives a slightly condensed form of the resulting frequency distributions.

Table C.1 Frequency of occurrences of 'kai' in 10 Pauline works

Number of sentences with	Romans (1–15)	1st Corinth.	2nd Corinth.	Galat.	Philip.
0 kai	386	424	192	128	42
1 kai	141	152	86	48	29
2 kais	34	35	28	5	19
3 or more kais	17	16	13	6	12
No. of sentences	578	627	319	187	102
Total number of kais	282	281	185	82	107

Number of sentences with	Colos.	1st Thesal.	1st Timothy	2nd Timothy	Hebrews
0 kai	23	34	49	45	155
1 kai	32	23	38	28	94
2 kais	17	8	9	11	37
3 or more kais	9	16	10	4	24
No. of sentences	81	81	106	88	310
Total number of kais	99	99	91	68	253

THE ANALYSIS

The analysis for this problem is nonstandard apart from the preliminary calculation of means and standard deviations. The summary statistics of Table C.2, *App. Stat.*, i.e. mean number of kais, modified mean (taking 3 or more kais equal to 3), modified standard deviation (using the modified data), ratio of variance to mean, and standard error of the mean, being simply transformations down each column of data, are easily obtained using CALCULATE commands.

The consistency within any proposed group of k means $\bar{Y}_1, \ldots, \bar{Y}_k$ with standard errors $\sqrt{v_1}, \ldots, \sqrt{v_k}$ is tested by the χ^2 statistic with $k - 1$ degrees of freedom

$$\chi^2 = \sum (\bar{Y}_j - \bar{Y}.)^2 / v_j \qquad \text{(C.1, App. Stat.)}$$

$$= \sum \bar{Y}_j^2 / v_j - \left(\sum \bar{Y}_j / v_j \right)^2 \left(\sum 1 / v_j \right)^{-1},$$

where $\bar{Y}. = (\sum \bar{Y}_j / v_j)(\sum 1 / v_j)^{-1}$.

High-quality graphics are used to illustrate the results; see Fig. C.1 after the Genstat program.

PROGRAM

```
 2   JOB        'Example C'
 3
 4   OPEN       'XC'; 2
 5
 6   "Data for frequencies
-7   386 424 192 128   42   23   34   49   45 155
-8   ...
-9    17  16  13   6  12    9  16  10   4  24 :
-10  followed by the total numbers of kais
-11  282 281 185 82 107 99 99 91 68 253:"
12
13   TEXT       NKAI; !T('0 kai','1 kai','2 kais','3 or more kais')
14   FACTOR     [LABELS=NKAI] KAI
15   TEXT       NWK; !T('Romans', '1st Cor','2nd Cor','Galat','Philip','Coloss',  \
16              '1st Thess','1st Tim','2nd Tim','Hebrews')
17   FACTOR     [LABELS=NWK] WORKS
18   TABLE      [CLASS=KAI,WORKS] SNTNCS
19   TABLE      [CLASS=WORKS] MMKAI,MSDKAI,MSEKAI,MVKAI,RATIO,RMKAI,RTKAI,TSNTN
20
21   READ       [CHANNEL=2; SERIAL=YES] SNTNCS,RTKAI
```

Identifier	Minimum	Mean	Maximum	Values	Missing	
SNTNCS	4.00	61.98	424.00	40	0	Skew
RTKAI	68.0	154.7	282.0	10	0	

```
22
23   "Frequency of occurrences of 'kai' in 10 Pauline works:"
24
25   PRINT      SNTNCS; FIELD=12; DEC=0                          "Table C.1 App. Stat."
```

	SNTNCS				
WORKS	Romans	1st Cor	2nd Cor	Galat	Philip
KAI					
0 kai	386	424	192	128	42
1 kai	141	152	86	48	29
2 kais	34	35	28	5	19
3 or more kais	17	16	13	6	12

WORKS	Coloss	1st Thess	1st Tim	2nd Tim	Hebrews
KAI					
0 kai	23	34	49	45	155
1 kai	32	23	38	28	94
2 kais	17	8	9	11	37
3 or more kais	9	16	10	4	24

```
  26    "To produce the summary statistics, a work table containing 10 sets of values
 -27     0, 1, 2, and 3 is required:"
  28
  29    TABLE      [CLASS=KAI,WORKS] DUMMY; !(10(0,1,2,3))
  30    CALCULATE  TSNTN = SNTNCS
  31    &          RMKAI = RTKAI/TSNTN
  32    &          MMKAI = SNTNCS*DUMMY/TSNTN
  33    &          MVKAI = SNTNCS*(DUMMY-MMKAI)**2/(TSNTN-1)
  34    &          MSDKAI = SQRT(MVKAI)
  35    &          RATIO = MVKAI/MMKAI
  36    &          MSEKAI = SQRT(MVKAI/TSNTN)
  37
  38    "Redefine NWK to display full names of works."
  39
  40    TEXT       NWK; !T('Romans (1-15)', '1st Corinthians','2nd Corinthians',      \
  41                       'Galatians','Philippians','Colossians','1st Thessalonians', \
  42                       '1st Timothy','2st Timothy','Hebrews')
  43
  44    "          Summary statistics from 10 works
 -45
 -46                Number of  Number   Mean no. Modified Modified Modified     mse
 -47                sentences  of kais  of kais  mean MM  st. dev. var/MM     (MM)"
  48
  49    PRINT      NWK,TSNTN,RTKAI,RMKAI,MMKAI,MSDKAI,RATIO,MSEKAI;                   \
  50               FIELD=18,2(8),3(10),9,10; DEC=0,0,0,3(4),2,4    "Table C.2 App. Stat."
```

NWK	TSNTN	RTKAI	RMKAI	MMKAI	MSDKAI	RATIO	MSEKAI
Romans (1-15)	578	282	0.4879	0.4498	0.7366	1.21	0.0306
1st Corinthians	627	281	0.4482	0.4306	0.7147	1.19	0.0285
2nd Corinthians	319	185	0.5799	0.5674	0.8171	1.18	0.0457
Galatians	187	82	0.4385	0.4064	0.6999	1.21	0.0512
Philippians	102	107	1.0490	1.0098	1.0388	1.07	0.1029
Colossians	81	99	1.2222	1.1481	0.9632	0.81	0.1070
1st Thessalonians	81	99	1.2222	1.0741	1.1487	1.23	0.1276
1st Timothy	106	91	0.8585	0.8113	0.9473	1.11	0.0920
2st Timothy	88	68	0.7727	0.7045	0.8598	1.05	0.0917
Hebrews	310	253	0.8161	0.7742	0.9386	1.14	0.0533

```
  51    "To reproduce Fig C.1, App. Stat.,  two lines are needed for each work:
 -52    (1) from (mm,mm) to (mm,0)  (2) from (mm-se,mm) to (mm+se,mm).
 -53    Form the sums and differences, and set up the Graphics environment:"
  54
  55    TABLE      [CLASS=WORKS] LOWER,UPPER
  56    CALCULATE  LOWER = RMKAI-MSEKAI  &  UPPER = RMKAI+MSEKAI
  57    OPEN       'C.PLT'; 6; Graphics
  58    DEVICE     6
  59    AXES       [EQUAL=SCALE] WINDOW=1; XTITLE='Mean numbers of kais';            \
  60               STYLE=X; YLOWER=0.0; YUPPER=4; XLOWER=0.3; XUPPER=1.4;            \
  61               XMARKS=!(0.4,0.6...1.2)
  62    PEN        2,3; LINESTYLE=2,1; METHOD=LINE
  63    TEXT       SYMBOLS[1...10]; '1','2','3','4','5','6','7','8','9','X'
  64
  65    " and then run through the ten works individually:"
  66
  67    FOR        W=1...10; H=1.0,1.5,0.5,0.5,0.5,1.0,1.5,1.0,0.5,1.5; SYMBOL=SYMBOLS[]
  68    SCALAR     M; #RMKAI$[W]
  69    VARIATE    X1; !(#LOWER$[W],#UPPER$[W])
  70    &          Y1; !(H,H)
  71    &          X2; !(M,M)
  72    &          Y2; !(0,H)
  73    PEN        2,3; SYMBOLS=SYMBOL,' '
  74    DGRAPH     [SCREEN=KEEP; KEYWINDOW=0] Y1,Y2; X1,X2; PEN=2,3
  75    ENDFOR
  76
```

```
 77   "Up to this point it has been simplest to keep the information in
-78    one-way tables classified by WORKS. To compute values for subsets
-79    of the data we must transfer the values to variates, since the RESTRICT
-80    directive only operates on vectors; and the factor must be given values:"
 81
 82   VARIATE    [10] VMEANS,VVARS,VSNTN,WEIGHTS
 83   EQUATE     MMKAI,MVKAI,TSNTN; VMEANS,VVARS,VSNTN
 84   CALCULATE  WEIGHTS = VSNTN/VVARS
 85   FACTOR     [MODIFY=YES]  WORKS; !(1...10)
 86
 87   "and we can now set up sets of proposed groupings of the works
-88    and display the group members and the corresponding value of chisquared."
 89
 90   VARIATE    GROUP[1...19]; !(1,2,3,4),!(1,2,4),!(1,3,4),!(1,2,3),!(2,3,4),!(1,3),\
 91   !(6,7),!(8,9,10),!(3,8,9,10),!(5,8,9,10),!(6,8,9,10),!(7,8,9,10),!(5,7,8,9,10),\
 92   !(6,7,8,9,10),!(5,6,8,9,10),!(5,6,7),!(5,6,7,8),!(5,6,7,9),!(5,6,7,10)
 93
 94   FOR        G = GROUP[]
 95   RESTRICT   NWK,VMEANS,WEIGHTS; WORKS.IN.G
 96   CALCULATE  CHISQ = SUM(WEIGHTS*VMEANS**2)-SUM(WEIGHTS*VMEANS)**2/SUM(WEIGHTS)
 97     &        DF = NVAL(G)-1
 98   SKIP       [FILETYPE=OUTPUT]    2
 99   PRINT      [ORIENT=ACROSS; IPRINT=*] NWK; FIELD=20; DECIMALS=0
100   PRINT      [SQUASH=Y; IPRINT=*] DF,CHISQ; FIELD=6,12; DECIMALS=0,2
101   ENDFOR                                                "Table C.3 App. Stat."


        Romans (1-15)     1st Corinthians    2nd Corinthians        Galatians

    3        7.70


102
103   "  ......  (We show the output for the first group only.)"
104
105   STOP

******** End of Example C.  Maximum of 17530 data units used at line 75 (32460 left)
```

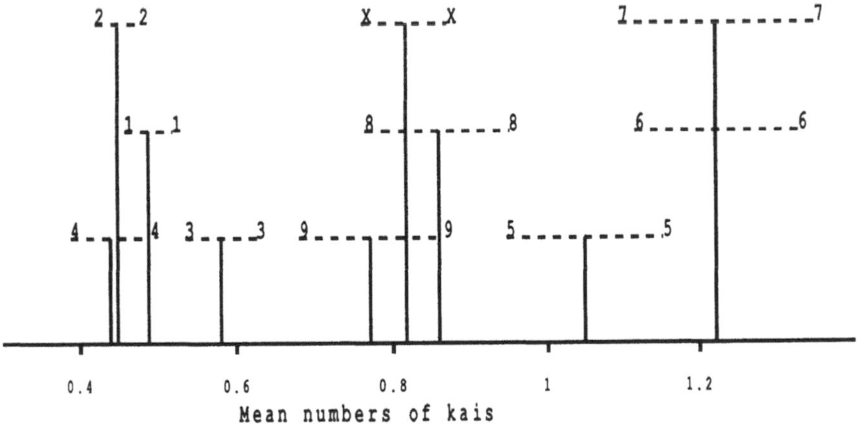

Figure C.1

SUGGESTED FURTHER WORK

Fit a censored negative binomial distribution by the method of maximum likelihood to each literary work. Compare the resulting analysis with that given here.

Example D

Temperature distribution in a chemical reactor

DESCRIPTION OF DATA*

A chemical reactor has 1250 sections and it is possible to calculate a theoretical temperature for each section. These have a distribution across sections with mean 452 °C and standard deviation 22 °C; the distribution is closely approximated by a normal distribution in the range 390–520 °C. For a variety of reasons, measured temperatures depart from the theoretical ones, the discrepancies being partly random and partly systematic. The temperature is measured in 20 sections and Table D.1 gives the measurements and the corresponding theoretical values. It is known further that the measured temperature in a section deviates from the 'true' temperature by a completely random error of measurement of zero mean and standard deviation 3 °C. Special interest

Table D.1 Measured and theoretical temperatures in 20 sections of reactor

Measured temp. (°C)	Theoretical temp. (°C)	Measured temp. (°C)	Theoretical temp. (°C)
431	432	472	498
450	470	465	451
431	442	421	409
453	439	452	462
481	502	451	491
449	445	430	416
441	455	458	481
476	464	446	421
460	458	466	470
483	511	476	477

* Fictitious data based on a real investigation.

attaches to the number of channels in a reactor with 'true' temperature above 490 °C.

THE ANALYSIS

The working assumption made in this problem is that over the reactor we can treat the true and measured temperatures T and T_{ME} as random with

$$T = \alpha + \beta(t_{TH} - t) + \varepsilon,$$
$$T_{ME} = T + \varepsilon',$$

(D.1, *App. Stat.*)

where t is the mean theoretical temperature in the measured channels, ε is a random term, and ε' is a measurement error independent of T.

Standard linear regression of measured temperature on theoretical temperature can be used to obtain an estimate $\hat{\beta}$ of the slope and hence of the mean of the true temperature, i.e. $\bar{T}_{ME} + \hat{\beta}(452 - \bar{t}_{TH})$ and standard deviation of this estimated mean, i.e. $\sqrt{(484\hat{\beta}^2 + s^2 - 9)}$, where s^2 is the residual mean square.

The proportion of values of T above 490 is estimated assuming normality and approximate confidence limits are calculated via large sample theory as discussed in *App. Stat.*

PROGRAM

```
 2  JOB        'Example D'
 3
 4  OPEN       'XD'; 2
 5
 6  "Data:
-7  431 432 472 498
-8  ...
-9  483 511 476 477 : "
10
11  UNITS      [20]
12  READ       [CHANNEL=2] MEASURED,THEORY
```

Identifier	Minimum	Mean	Maximum	Values	Missing
MEASURED	421.0	454.6	483.0	20	0
THEORY	409.0	459.7	511.0	20	0

```
13
14  MODEL      MEASURED
15  FIT        THEORY
```

***** Regression Analysis *****

Response variate: MEASURED
    Fitted terms: Constant, THEORY

*** Summary of analysis ***

	d.f.	s.s.	m.s.	v.r.
Regression	1	4075.	4075.2	34.80
Residual	18	2108.	117.1	
Total	19	6183.	325.4	

Percentage variance accounted for 64.0

* MESSAGE: The following units have high leverage:
```
           6        0.214
          19        0.218
```

*** Estimates of regression coefficients ***

```
                    estimate        s.e.          t
Constant              220.1         39.8        5.53
THEORY               0.5101       0.0865        5.90
```

```
16    " and this gives us all that we really need. However, various quantities
-17   computed while producing the regression analysis can be extracted for
-18   further use by using the RKEEP directive."
19
20   RKEEP      FITTEDVALUES=FV; ESTIMATES=ESTS; SE=ERRS; DEVIANCE=RSS; DF=DF
21
22   "It is usually sensible to display the fitted line and the observed values."
23
24   GRAPH      [NROWS=22] FV,MEASURED;THEORY;   METHOD=LINE,POINT
```

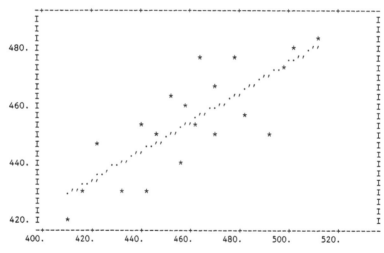

FV   v. THEORY using symbol .
MEASURED v. THEORY using symbol *

```
25   "The estimates (ESTS) and errors (ERRS) will be variates long enough to contain
-26  values for all the parameters, including the constant term. The second elements
-27  of these variates will be the figures relating to the fitted term THEORY."
28
29   CALCULATE SLOPE,ESE = (ESTS,ERRS)$[2]
30
31   "and the various quantities discussed in App. Stat. can be evaluated using
-32  Genstat functions SQRT, MEAN, NVAL and NORMAL:"
33
34   CALCULATE RMS = RSS/DF
35   &         ESD = SQRT(RMS)
36   &         SM  = MEAN(MEASURED)
37   &         SEM = SQRT(RMS/NVAL(MEASURED))
38   &         TTM = SM+SLOPE*(452-MEAN(THEORY))
39   &         SDT = SQRT(484*SLOPE**2+RMS-9)
40   &         ZETA = (490-TTM)/SDT
41   &         PHI = 1250*NORMAL(-ZETA)
42
```

```
43  PRINT      SLOPE,ESE,ESD,SM,SEM,TTM,SDT,ZETA,PHI

      SLOPE         ESE         ESD          SM         SEM         TTM         SDT
     0.5101     0.08647       10.82       454.6       2.420       450.7       15.30


       ZETA         PHI
      2.571       6.343

44  STOP

******** End of Example D.  Maximum of 8334 data units used at line 43 (41026 left)
```

Example E

A 'before and after' study of blood pressure

DESCRIPTION OF DATA

Table E.1 gives, for 15 patients with moderate essential hypertension, supine systolic and diastolic blood pressures immediately before and two hours after taking 25 mg of the drug captopril. The data were provided by Dr G. A. MacGregor, Charing Cross Hospital Medical School; for a report on the investigation and appreciable further summary data, see MacGregor, Markandu, Roulston and Jones (1979).

Table E.1 Blood pressures (mm Hg) before and after captopril

Patient no.	Systolic			Diastolic		
	before	after	difference	before	after	difference
1	210	201	−9	130	125	−5
2	169	165	−4	122	121	−1
3	187	166	−21	124	121	−3
4	160	157	−3	104	106	2
5	167	147	−20	112	101	−11
6	176	145	−31	101	85	−16
7	185	168	−17	121	98	−23
8	206	180	−26	124	105	−19
9	173	147	−26	115	103	−12
10	146	136	−10	102	98	−4
11	174	151	−23	98	90	−8
12	201	168	−33	119	98	−21
13	198	179	−19	106	110	4
14	148	129	−19	107	103	−4
15	154	131	−23	100	82	−18

THE ANALYSIS

The main interest here lies in computing a measure of treatment effect and plotting the data. Differences in systolic and diastolic blood pressure, 'after' minus 'before', are treated as response measures and are calculated within Genstat.

For a check on whether the treatment effect differs according to initial value of blood pressure, the above differences can be plotted against the 'before' readings. If a regression line is fitted (see Example D for regression using Genstat) it should be interpreted cautiously as both axes represent random variables; see the detailed discussion in *App. Stat.* However, for the data of this example no evidence of such a relationship exists for either the systolic or diastolic response and so we do not trouble to display these plots here. Instead we inspect simply the relationship between systolic and diastolic differences using a line-printer plot.

PROGRAM

```
 2  JOB        'Example E'
 3
 4  OPEN       'XE'; 2
 5
 6  "Data:
-7  210 201 130 125
-8  ...
-9  154 131 100  82 : "
10
11  UNITS      [15]
12  READ       [CHANNEL=2] SYSBEF,SYSAFT,DIABEF,DIAAFT

    Identifier  Minimum     Mean    Maximum    Values   Missing
       SYSBEF     146.0     176.9      210.0       15         0
       SYSAFT     129.0     158.0      201.0       15         0
       DIABEF      98.0     112.3      130.0       15         0
       DIAAFT      82.0     103.1      125.0       15         0
13
14  "Form differences and simple statistics."
15
16  CALCULATE  SYSDIF,DIADIF = SYSAFT,DIAAFT-SYSBEF,DIABEF
17    &        MSD,MDD = MEAN(SYSDIF,DIADIF)
18    &        SSD,DSD = SQRT(VAR(SYSDIF,DIADIF))
19    &        SSE,DSE = SSD,DSD/SQRT(15)
20  PRINT      MSD,MDD,SSD,DSD,SSE,DSE; DECIMALS=2          "Table E.2, App.Stat."

         MSD         MDD         SSD         DSD         SSE         DSE
      -18.93       -9.27        9.03        8.61        2.33        2.22

21  FACTOR     [LEVELS=15] PATIENT; !(1...15)
22  GRAPH      [TITLE=!T('Systolic difference (''after'' - ''before'')\
-23   versus diastolic difference');                                    \
24             YTITLE=!T('Systolic difference');                        \
25             XTITLE=!T('Diastolic difference'); NROWS=25]             \
26             SYSDIF; DIADIF; SYMBOLS=PATIENT          "Fig E.1 App. Stat."
```

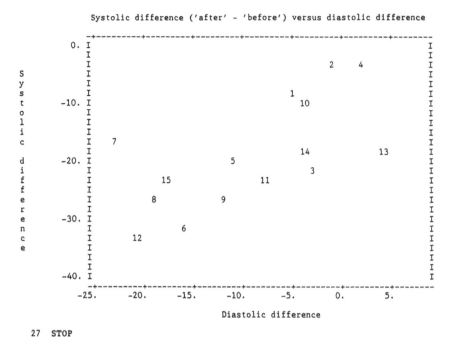

Systolic difference ('after' - 'before') versus diastolic difference

27 STOP

******** End of Example E. Maximum of 7702 data units used at line 26 (42288 left)

SUGGESTED FURTHER WORK

1. Repeat the analysis using log blood pressures. How would you decide (perhaps given more extensive data) which analysis is to be preferred?
2. Suppose it is suggested that log(diastolic × systolic) and log(diastolic/ systolic) are good indices of overall blood pressure and blood pressure profile, respectively. How would you proceed?

Example F

Comparison of industrial processes in the presence of trend

DESCRIPTION OF DATA*

In a plant-scale experiment on the production of a certain chemical, a batch of intermediate product was divided into six equal portions which were then processed on successive days by two different methods, P_1 and P_2. The order of treatment and the yields are given in Table F.1. It was expected that superimposed on any process effect there would be a smooth, roughly parabolic trend. Experience of similar experiments showed that the standard deviation of a single observation was about 0.1.

Table F.1 Treatment and yields in plant-scale experiment

Day	1	2	3	4	5	6
Process	P_1	P_2	P_2	P_1	P_1	P_2
Yield	5.84	5.73	7.30	10.46	9.71	5.91

THE ANALYSIS

We could go ahead simply regressing yield on process plus a parabolic trend by fitting the model

$$E(Y) = \begin{cases} \mu + \beta t + \gamma t^2, & \text{Process 1} \\ \mu + \beta t + \gamma t^2 + \tau, & \text{Process 2} \end{cases}$$

where $t = 1, \ldots, 6$ denotes day. There can however be an advantage to

* Fictitious data based on a real investigation.

using orthogonal polynomials to represent the quadratic trend, particularly if the time delay to maximum yield, and its standard error, are to be estimated. In this case we write the model as

$$E(Y) = \mu \pm \tau + \beta_1\phi_1 + \beta_2\phi_2,$$

where ϕ_1 and ϕ_2 denote linear and quadratic orthogonal polynomials; ϕ_1 and ϕ_2 are computed using the procedure ORTHPOL although with so few lines of data we could equally well have chosen to read in the values

$$\phi_1 = -5, -3, -1, 1, 3, 5 \text{ and}$$
$$\phi_2 = 5, -1, -4, -4, -1, 5$$

(Pearson and Hartley, 1966, Table 47).

An approximate variance for the estimated time delay corresponding to maximum yield is given by substituting estimated values into the expression

$$\tfrac{4}{9}\{\beta_2^2 \text{ var } (\hat{\beta}_1) + \beta_1^2 \text{ var } (\hat{\beta}_2)\}/\beta_2^4 \quad \text{(F.4, App. Stat.)}$$

PROGRAM

```
 2  JOB        'Example F'
 3
 4  OPEN       'XF'; 2
 5
 6  "Data:
-7  5.84  5.73  7.30  10.46  9.71  5.91 : "
 8
 9  UNITS      [6]
10  FACTOR     [LABELS=!T(P1,P2)] PROCESS; !(1,2,2,1,1,2)
11  VARIATE    DAY; !(1...6)
12
13  READ       [CHANNEL=2] YIELD

    Identifier  Minimum    Mean   Maximum    Values   Missing
         YIELD    5.730    7.492   10.460         6         0
14
15  "The data can be displayed with default option settings:
-16  GRAPH      YIELD; DAY; SYMBOLS=PROCESS
-17  or for a better display:"                              "Fig F.1 App. Stat."
18
19  GRAPH      [TITLE='Yields in plant-scale experiment for processes P1 and P2';  \
20             YLOWER=5; YUPPER=11; XLOWER=0; XUPPER=7; XINTEGER=Y; YINTEGER=Y;     \
21             NROW=20; NCOLUMNS=70; XTITLE='Day']   YIELD; DAY; SYMBOLS=PROCESS
```

Yields in plant-scale experiment for processes P1 and P2

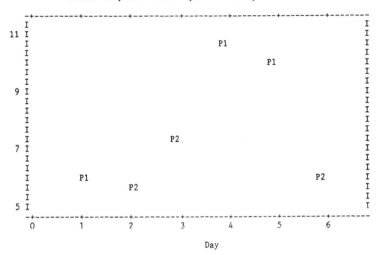

Day

```
  22   ORTHPOL   [MAX=2] DAY; D
  23
  24   "This standard Genstat procedure produces orthogonal polynomials which differ
 -25   by constant factors from those used in App. Stat.; Pearson and Hartley tabulate
 -26   integral values which are more convenient for hand calculation. To allow
 -27   direct comparison of the estimates:"
  28
  29   CALCULATE D[ ] = 2,1.5 * D[ ]
  30
  31   MODEL      YIELD
  32   TERMS      YIELD,PROCESS,D[ ]
  33   FIT        PROCESS,D[ ]
```

***** Regression Analysis *****

Response variate: YIELD
    Fitted terms: Constant, PROCESS, D[1], D[2]

*** Summary of analysis ***

	d.f.	s.s.	m.s.	v.r.
Regression	3	22.05825	7.35275	337.02
Residual	2	0.04363	0.02182	
Total	5	22.10188	4.42038	
Change	-3	-22.05825	7.35275	337.02

Percentage variance accounted for 99.5

*** Estimates of regression coefficients ***

	estimate	s.e.	t
Constant	8.7556	0.0855	102.43
PROCESS P2	-2.528	0.121	-20.86
D[1]	0.2568	0.0177	14.48
D[2]	-0.3301	0.0161	-20.48

```
  34   "Note that the s.e.s are based on the calculated r.m.s. 0.0218, not on
 -35   an assumed 0.01. MODEL [DISPERSION=0.01] YIELD could be used to obtain
 -36   the values given in App. Stat.
 -37
```

```
-38   The simplest way to get an idea of the position of maximal yield is
-39   to plot the fitted values after dropping Process:"
 40
 41   DROP      [PRINT=*] PROCESS
 42   RKEEP     YIELD; FITTED=FV
 43   GRAPH     [NROW=20] FV; DAY; METHOD=CURVE; SYMBOL='*'
```

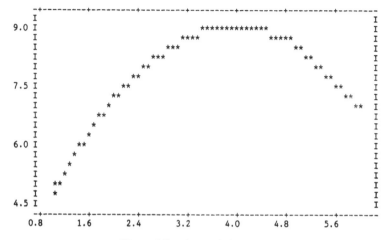

FV v. DAY using symbol *

```
 44   "which suggests a peak at about 4 days."
 45
 46   "To do any better, we have to discover the exact forms of polynomials
-47   produced by ORTHPOL. By printing the values of D[1] and D[2] it can
-48   be seen that (for n=6, after adjustment)
-49
-50       phi1(x):   2*x - 7
-51       phi2(x):   (3*phi1(x)**2 - 35)/8 = (3*x**2-21*x)/2
-52
-53   and from the fitted equation for process 2,
-54
-55       y = 8.7556 - 2.528 + 0.2568*phi1(x) - 0.3301*phi2(x)
-56
-57       with a maximum at x = 3.97965/2*0.49515 = 4.02  "
 58
 59   STOP

******** End of Example F.  Maximum of 10406 data units used at line 32 (38954 left)
```

SUGGESTED FURTHER WORK

1. Repeat the analysis taking the model in the forms
   (a) $\mu \pm \tau + \beta t + \gamma t^2$
   (b) $\mu + \beta t + \gamma t^2$ (Process 1), $\mu + \tau + \beta t + \gamma t^2$ (Process 2).
2. Make a careful comparison of the answers from (a) and (b) with those obtained previously.

Example G

Cost of construction of nuclear power plants

DESCRIPTION OF DATA

Table G.1 gives data, reproduced by permission of the Rand Corporation, from a report (Mooz, 1978) on 32 light water reactor (LWR) power plants constructed in the USA. It is required to predict the capital cost involved in the construction of further LWR power plants. The notation used in Table G.1 is explained in Table G.2. The final six lines of data in Table G.1 relate to power plants for which there were partial turnkey guarantees and for which it is possible that some manufacturers' subsidies might be hidden in the quoted capital costs.

Table G.1 Data on 32 LWR power plants in the USA

C	D	T_1	T_2	S	PR	NE	CT	BW	N	PT
460.05	68.58	14	46	687	0	1	0	0	14	0
452.99	67.33	10	73	1065	0	0	1	0	1	0
443.22	67.33	10	85	1065	1	0	1	0	1	0
652.32	68.00	⟨11	67	1065	0	1	1	0	12	0
642.23	68.00	11	78	1065	1	1	1	0	12	0
345.39	67.92	13	51	514	0	1	1	0	3	0
272.37	68.17	12	50	822	0	0	0	0	5	0
317.21	68.42	14	59	457	0	0	0	0	1	0
457.12	68.42	15	55	822	1	0	0	0	5	0
690.19	68.33	12	71	792	0	1	1	1	2	0
350.63	68.58	12	64	560	0	0	0	0	3	0
402.59	68.75	13	47	790	0	1	0	0	6	0
412.18	68.42	15	62	530	0	0	1	0	2	0
495.58	68.92	17	52	1050	0	0	0	0	7	0
394.36	68.92	13	65	850	0	0	0	1	16	0
423.32	68.42	11	67	778	0	0	0	0	3	0
712.27	69.50	18	60	845	0	1	0	0	17	0

Table G.1 (*cont.*)

C	D	T_1	T_2	S	PR	NE	CT	BW	N	PT
289.66	68.42	15	76	530	1	0	1	0	2	0
881.24	69.17	15	67	1090	0	0	0	0	1	0
490.88	68.92	16	59	1050	1	0	0	0	8	0
567.79	68.75	11	70	913	0	0	1	1	15	0
665.99	70.92	22	57	828	1	1	0	0	20	0
621.45	69.67	16	59	786	0	0	1	0	18	0
608.80	70.08	19	58	821	1	0	0	0	3	0
473.64	70.42	19	44	538	0	0	1	0	19	0
697.14	71.08	20	57	1130	0	0	1	0	21	0
207.51	67.25	13	63	745	0	0	0	0	8	1
288.48	67.17	9	48	821	0	0	1	0	7	1
284.88	67.83	12	63	886	0	0	0	1	11	1
280.36	67.83	12	71	886	1	0	0	1	11	1
217.38	67.25	13	72	745	1	0	0	0	8	1
270.71	67.83	7	80	886	1	0	0	1	11	1

Table G.2 Notation for data of Table G.1

C	Cost in dollars $\times 10^{-6}$, adjusted to 1976 base
D	Date construction permit issued
T_1	Time between application for and issue of permit
T_2	Time between issue of operating license and construction permit
S	Power plant net capacity (MWe)
PR	Prior existence of an LWR on same site ($= 1$)
NE	Plant constructed in north-east region of USA ($= 1$)
CT	Use of cooling tower ($= 1$)
BW	Nuclear steam supply system manufactured by Babcock-Wilcox ($= 1$)
N	Cumulative number of power plants constructed by each architect-engineer
PT	Partial turnkey plant ($= 1$)

THE ANALYSIS

This is an interesting problem in that a number of models may be found to fit the data almost equally well. Table G.4 *App. Stat.* gives a six-variable model, with log C regressed on PT, CT, log N, log S, D, NE, this model being selected by successive elimination. Interactions with PT, a variable identifying a subgroup of the observations, are examined but none is found to be significant.

Other models, not considered in *App. Stat.*, include the possibility of a $(\log N)^2$ term. Also, Ling (1984) in a review of *App. Stat.* suggests there is evidence for a different model when $N > 12$ and so defines an indicator variable $I = 0$ for $N \leq 12$, $I = 1$ for $N > 12$, and fits three models containing the sets:

1. CT, *T*, *S*, *D*, NE, *D* × *N*;
2. CT, *I*, *S*, *D*, NE, *D* × *N*, *N*; and
3. CT, *I*, log *S*, *D*, NE, *D* × log *N*, log *N*.

The adequacy of all competing models should be examined, the aim being to find a model which is a good predictor of capital cost. Where detailed interpretation of the estimated parameters and of the choice of variables is intrinsically important, special care is necessary.

We choose here to fit the full model containing all ten explanatory variables (as in *App. Stat.*) and then remove the least significant terms successively using the DROP command. Alternatively we could use the STEP command, as in Example P, for automatic backward or forward selection. In general any automatic procedure must be used with caution to arrive at an appropriate selection, although here it would lead either way to the six-variable model of Table G.4, *App. Stat.*

PROGRAM

```
 2  JOB       'Example G'
 3
 4  OPEN      'XG' ; 2
 5
 6  "Data:
-7  460.05  68.58  14  46     687  0  1  0  0 14  0
-8  ...
-9  270.71  67.83   7  80     886  1  0  0  1 11  1 : "
10
11  UNITS     [32]
12  FACTOR    [LEVELS=!(0,1)] PR,NE,CT,BW,PT
13
14  "These structures could be declared as variates, but they are better treated
-15  as factors and interactions are handled more easily."
16
17  READ      [CHANNEL=2] C,D,T1,T2,S,PR,NE,CT,BW,N,PT
```

Identifier	Minimum	Mean	Maximum	Values	Missing
C	207.5	461.6	881.2	32	0
D	67.17	68.58	71.08	32	0
T1	7.00	13.75	22.00	32	0
T2	44.00	62.38	85.00	32	0
S	457.0	825.4	1130.0	32	0
N	1.000	8.531	21.000	32	0

```
18
19  CALCULATE lnC,lnT1,lnT2,lnS,lnN = LOG(C,T1,T2,S,N)
20
21  "The independent variable is given in a MODEL statement, and the full set
-22  of variables involved (dependent and independent) named in a TERMS statement.
-23  This sets up a complete sums and squares matrix which will be used by all
-24  regression commands until another TERMS statement is given."
25
26  MODEL     lnC
27  TERMS     lnC,D,lnT1,lnT2,lnS,PR,NE,CT,BW,lnN,PT
28
29  "First, fit the full model."
```

```
   30
   31  FIT       D,lnT1,lnT2,lnS,PR,NE,CT,BW,lnN,PT
```

***** Regression Analysis *****

```
  Response variate: lnC
       Fitted terms: Constant, D, lnT1, lnT2, lnS, PR, NE, CT, BW, lnN, PT
```

*** Summary of analysis ***

	d.f.	s.s.	m.s.	v.r.
Regression	10	3.8600	0.38600	14.27
Residual	21	0.5680	0.02705	
Total	31	4.4281	0.14284	
Change	-10	-3.8600	0.38600	14.27

Percentage variance accounted for 81.1

```
  * MESSAGE: The following units have large standardized residuals:
               26      -2.19
```

*** Estimates of regression coefficients ***

	estimate	s.e.	t
Constant	-14.24	4.23	-3.37
D	0.2092	0.0653	3.21
lnT1	0.092	0.244	0.38
lnT2	0.286	0.273	1.05
lnS	0.694	0.136	5.10
PR 1	-0.0924	0.0773	-1.19
NE 1	0.2581	0.0769	3.35
CT 1	0.1204	0.0663	1.82
BW 1	0.033	0.101	0.33
lnN	-0.0802	0.0460	-1.74
PT 1	-0.224	0.122	-1.83

```
   32  "Note the warning message indicating the unit(s) with large residuals.
  -33  We could now remove the less significant terms successively by
  -34  DROP       [PRINT=M,S]    BW & lnT1 & lnT2
  -35  The output would include analysis of variance tables and messages about
  -36  units with large residuals or leverage. To fit the six term model of
  -37  Table G.4, App. Stat. we remove also PR:"
   38
   39  DROP       [PRINT=M,S,E]  BW,lnT1,lnT2,PR
```

***** Regression Analysis *****

```
  Response variate: lnC
       Fitted terms: Constant, D, lnS, NE, CT, lnN, PT
```

*** Summary of analysis ***

	d.f.	s.s.	m.s.	v.r.
Regression	6	3.7943	0.63239	24.95
Residual	25	0.6337	0.02535	
Total	31	4.4281	0.14284	
Change	4	0.0657	0.01643	0.65

Percentage variance accounted for 82.3

```
  * MESSAGE: The following units have large standardized residuals:
                7      -2.22
               19       2.28
```

*** Estimates of regression coefficients ***

	estimate	s.e.	t
Constant	-13.26	3.14	-4.22
D	0.2124	0.0433	4.91
lnS	0.723	0.119	6.09
NE 1	0.2490	0.0741	3.36
CT 1	0.1404	0.0604	2.32
lnN	-0.0876	0.0415	-2.11
PT 1	-0.226	0.114	-1.99

```
  40  RKEEP     lnC; RESIDUALS=R; FITTED=F
  41
  42  "The differences calculated by Genstat are standardised; they are
 -43  scale-free and have constant standard error. Alternatively nonstandardised
 -44  residuals as used in App. Stat. can simply be calculated as the difference
 -45  between observed and fitted."
  46
  47  GRAPH     [YTITLE='Standardised Residuals, six variable model';          \
  48            XTITLE='Date construction permit issued'] R; D   "Fig G.1. App. Stat."
```

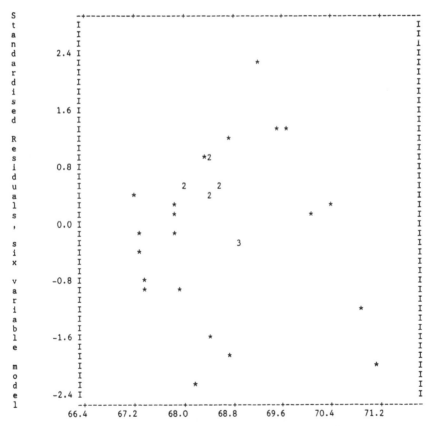

Date construction permit issued

```
  49  "The integers in the graph are counts of coincident points.
 -50
 -51  For plots of residuals against variables, e.g. lnS:
 -52  GRAPH     [XTITLE='ln (power plant net capacity)'] R; lnS
```

```
-53
-54  To get a graph of residuals against normal order statistics a standard
-55  Genstat procedure RCHECK may be used:"
 56
 57  RCHECK    [RESIDUAL=STANDARDISED] RESIDUAL; NORMAL         "Fig G.4. App. Stat"
```

```
     -+---------+---------+---------+---------+---------+---------+--------
    I I                                                                    I I
    I I                                                                    I I
    I I                                                                    I I
2.4 I I                                                                    I I
    I I                                                     *              I I
    I I                                                                    I I
    I I                                                                    I I
    I I                                                                    I I
    I I                                                                    I I
1.6 I I                                                                    I I
    I I                                             *   *                  I I
    I I                                          *                         I I
    I I                                                                    I I
    I I                                    *  **                           I I
0.8 I I                                                                    I I
    I I                               ****                                 I I
    I I                            ***                                     I I
    I I                          **                                        I I
    I I                        **                                          I I
0.0 I I                     **                                             I I
    I I                 ** *                                               I I
    I I               *                                                    I I
    I I                                                                    I I
    I I                                                                    I I
-0.8 I I             *                                                     I I
    I I          *  *                                                      I I
    I I         *                                                          I I
    I I                                                                    I I
    I I                                                                    I I
-1.6 I I      *                                                            I I
    I I    *                                                               I I
    I I   *                                                                I I
    I I                                                                    I I
    I I *                                                                  I I
-2.4 I I                                                                   I I
     -+---------+---------+---------+---------+---------+---------+--------
     -2.4     -1.6      -0.8      0.0       0.8       1.6       2.4
```

resid v. nquant using symbol *

```
 58  "We could redefine the maximal model as in App. Stat. to include only the six
-59  variables of interest and the interactions of 'partial turnkey plant' with
-60  Date (D), Cooling Tower (CT) and the (natural) logarithms of Capacity (s)
-61  and Engineer experience (N) by writing
-62  TERMS     lnC+PT*D+PT*lnS+NE+PT*CT+PT*lnN
-63  Note that in a model formula, a 'product' term X*Y is equivalent to
-64       (main effect) X + (main effect) Y + (interaction) X.Y
-65  After fitting main effects by
-66  FIT       [PRINT=*] PT,D,lnS,CT,lnN
-67  the effects of the interactions with PT can be examined by
-68  TRY       PT.(D+lnS+CT+lnN)
-69  The output is not shown here."
 70
 71  STOP

******** End of Example G.  Maximum of 26660 data units used at line 57 (23330 left)
```

SUGGESTED FURTHER WORK

Fit other models as discussed under 'THE ANALYSIS' heading. Examine their adequacy, particularly compared with the six-variable model fitted here.

Example H

Effect of process and purity index on fault occurrence

DESCRIPTION OF DATA*

Minor faults occur irregularly in an industrial process and, as an aid to their diagnosis, the following experiment was done. Batches of raw material were selected and each batch was divided into two equal sections: for each batch, one of the sections were processed by the standard method and the other by a slightly modified process, in which the temperature at one stage is reduced. Before processing, a purity index was measured for the whole batch of material. For the product from each section of material, it was recorded whether the minor faults did or did not occur. Results for 22 batches are given in Table H.1.

Table H.1 Occurrence of faults in 22 batches

Purity index	Standard process	Modified process	Purity index	Standard process	Modified process
7.2	NF	NF	6.5	NF	F
6.3	F	NF	4.9	F	F
8.5	F	NF	5.3	F	NF
7.1	NF	F	7.1	NF	F
8.2	F	NF	8.4	F	NF
4.6	F	NF	8.5	NF	F
8.5	NF	NF	6.6	F	NF
6.9	F	F	9.1	NF	NF
8.0	NF	NF	7.1	F	NF
8.0	F	NF	7.5	NF	F
9.1	NF	NF	8.3	NF	NF

F, faults occur.   NF, no faults occur.

*Fictitious data based on a real investigation.

THE ANALYSIS

Preliminary analysis of the occurrences of F and NF within the matched pairs, given the small number of observations, is best done by hand. Then a logistic model taking account of purity index and process can be fitted. The model is assumed to be

$$\text{pr}(\text{fault}|x_i) = \frac{\exp(\alpha + \beta x_i + \Delta)}{1 + \exp(\alpha + \beta x_i + \Delta)}$$

with $\Delta = 0$ for the standard process; x_i denotes the purity index.

High-quality graphics are used to illustrate the fitted model; see Fig. H.1 after the Genstat program.

PROGRAM

```
  2   JOB        'Example H'
  3
  4   OPEN       'XH'; 2
  5
  6   "The data are in the form shown in App. Stat., a purity index followed by
 -7   the result - F(ault) or N(ofault) - for the standard and modified processes
 -8   in 22 batches:
 -9   7.2 N N
-10   ...
-11   8.3 N N : "
 12
 13   UNITS      [22]
 14   TEXT       NF; !T(N,F)
 15   FACTOR     [LEVELS=2; LABELS=NF] FS,FM
 16   READ       [CHANNEL=2] P,FS,FM; FREP=LABELS
```

Identifier	Minimum	Mean	Maximum	Values	Missing
P	4.600	7.350	9.100	22	0

```
 17
 18   "To analyse the data, we need to construct structures with 44 values, one
-19   for each section; a purity index with the 22 listed values repeated, a
-20   factor indicating whether the process was 'standard' or 'modified'; and
-21   a response variate with vaules 0 (no fault) or 1 (fault)."
 22
 23   UNITS      [44]
 24   VARIATE    PURITY; !((#P)2)
 25   TEXT       NP; !T(standard,modified)
 26   FACTOR     [LEVELS=2; LABELS=NP] PROCESS; !(22(1,2))
 27   VARIATE    FAULTS; !(#FS,#FM)
 28
 29   "This assigns vaules 1 and 2, the default numerical levels for the factors."
 30
 31   CALCULATE FAULTS = FAULTS-1
 32
 33   "We can now fit a generalised linear model to the binomial variate FAULTS;
-34   the distribution must be given (as an option) and the totals, in this case 1
-35   for each unit, given as a parameter."
 36
 37   MODEL      [DISTRIBUTION=BINOMIAL; LINK=LOGIT] FAULTS; NBINOMIAL=!(44(1))
 38   TERMS      PURITY,PROCESS
 39
 40   "The commands
-41   FIT        [PRINT=M,S,E,A]  PURITY & PROCESS
-42   will fit logistic models (since a logit link was specified above) to
-43   (i) purity and (ii) process: we illustrate here the output from fitting
-44   (iii) purity and process:"
 45
 46   FIT        [PRINT=M,S,E,A]  PURITY,PROCESS
```

***** Regression Analysis *****

Response variate: FAULTS
  Binomial totals: !(1,...,1)
     Distribution: Binomial
    Link function: Logit
     Fitted terms: Constant, PURITY, PROCESS

*** Summary of analysis ***
    Dispersion parameter is 1

	d.f.	deviance	mean deviance	deviance ratio
Regression	2	6.72	3.361	2.61
Residual	41	52.81	1.288	
Total	43	59.53	1.385	
Change	-2	-6.72	3.361	2.61

* MESSAGE: The following units have high leverage:
                28   0.158

*** Estimates of regression coefficients ***

	estimate	s.e.	t
Constant	4.46	2.15	2.08
PURITY	-0.604	0.282	-2.14
PROCESS modified			
	-0.864	0.669	-1.29

* MESSAGE: s.e.s are based on dispersion parameter with value 1

*** Accumulated analysis of deviance ***

Change	d.f.	deviance	mean deviance	deviance ratio
+ PURITY				
+ PROCESS	2	6.722	3.361	2.61
Residual	41	52.812	1.288	
Total	43	59.534	1.385	

```
  47      "The estimates given are equivalent to those shown in App. Stat. The
 -48      parameterisation is different: when constants are fitted for the levels of a
 -49      factor, that for the first level is constrained to be zero. This avoids a known
 -50      linear dependency which would cause the fitting process to fail. (Unexpected
 -51      linear dependencies, arising from actual data values, may still be present -
 -52      these will be detected.) Ignoring the logistic part, the process fits
 -53
 -54         Mean              for the standard level of PROCESS
 -55      and Mean + Constant  for the modified level of PROCESS.
 -56
 -57      To display the observed and fitted values as in Fig. H.1 App. Stat., we use
 -58      the RKEEP directive to extract the fitted values from the current model:"
  59
  60      RKEEP     FAULTS; FITTED=FV
  61
  62      "Group the Purity Index suitably to form a factor FP and calculate the
 -63      mean index for each level of this new factor:"
  64
  65      SORT      [INDEX=PURITY;GROUPS=FP;LEVELS=LP;LIMITS=!(5.7,6.7,7.7,8.7)]
  66      TABULATE  [CLASS=PROCESS,FP] FAULTS; MEANS=TFAULTS
  67      VARIATE   [NV=10] PI    &  LL;  !((#LP)2)
  68      EQUATE    TFAULTS; PI
  69      TEXT      NPR; !T(S,M)
  70      FACTOR    [LABELS=NPR] SF; !(5(1,2))
  71
  72      OPEN      'H.PLT'; 6; Graphics
  73      DEVICE    6
  74      AXES      WINDOW=1; XTITLE='Purity index'; YTITLE='Proportion of faults';   \
  75                YLOWER=0.0; YUPPER=1.1; XLOWER=4.0; XUPPER=10.0; XINTEGER=Y;       \
```

```
76                XMARKS=!(5,6,7,8,9); YMARKS=!(0.2,0.4...1.0)
77  PEN           1,2,3; LINESTYLE=0,1,2; METHOD=POINT,LINE,LINE; SYMBOLS=SF,0,0
78  DGRAPH        [TITLE=' Observed faults (S,M) and fitted logistic models';        \
79                KEYWINDOW=0]  PI; LL; PEN=1
80  RESTRICT      FV,PURITY; PROCESS.EQ.1
81  DGRAPH        [SCREEN=KEEP;KEYWINDOW=0] FV; PURITY; PEN=2
82  RESTRICT      FV,PURITY; PROCESS.EQ.2
83  DGRAPH        [SCREEN=KEEP;KEYWINDOW=0] FV; PURITY; PEN=3
84
85  STOP
```

******** End of Example H. Maximum of 16076 data units used at line 80 (33914 left)

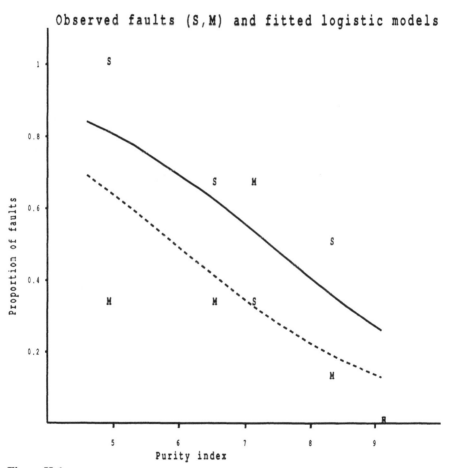

Figure H.1

SUGGESTED FURTHER WORK

1. Examine the effect of the observation shown to have high leverage.
2. Fit reduced models having $\Delta = 0$ or $\beta = 0$ or $\beta = \Delta = 0$ and interpret the results.

Example I

Growth of bones from chick embryos

DESCRIPTION OF DATA

Table I.1 gives data on the growth of bones from seven-day-old chick embryos after cultivation over a nutrient chemical medium (Biggers and Heyner, 1961). The observations are of log dry weight (μg). Two bones were available from each embryo and the experiment was therefore set out in a (balanced) incomplete block design with two units per block. C denotes the complete medium with about 30 ingredients in carefully controlled quantities. The five other media were obtained by omitting a single amino-acid, e.g., His$^-$ is a medium without L-histidine, etc. The treatment pairs were randomized, but the results are given in systematic order. Interest lies in comparing the effects of omitting the various amino-acids.

Table I.1 Log_{10}(dry weight) of tibiotarsi from seven-day-old chick embryos

Embryo							
1 C	2.51:	His$^-$	2.15	9 His$^-$	2.32:	Lys$^-$	2.53
2 C	2.49:	Arg$^-$	2.23	10 Arg$^-$	2.15:	Thr$^-$	2.23
3 C	2.54:	Thr$^-$	2.26	11 Arg$^-$	2.34:	Val$^-$	2.15
4 C	2.58:	Val$^-$	2.15	12 Arg$^-$	2.30:	Lys$^-$	2.49
5 C	2.65:	Lys$^-$	2.41	13 Thr$^-$	2.20:	Val$^-$	2.18
6 His$^-$	2.11:	Arg$^-$	1.90	14 Thr$^-$	2.26:	Lys$^-$	2.43
7 His$^-$	2.28:	Thr$^-$	2.11	15 Val$^-$	2.28:	Lys$^-$	2.56
8 His$^-$	2.15:	Val$^-$	1.70				

THE ANALYSIS

The within-block analysis in *App. Stat.* proceeds from first principles by setting up a regression model parameterized so as to estimate directly each effect of interest, i.e. the reduction in log_{10}(dry weight) due to omission of an amino-acid. A between-block estimate of the effects is obtained by a regression analysis applied to embryo totals and the two estimates for each effect are pooled by weighting according to the within- and between-variation.

Here instead we simply use the procedure ANOVA to do both the within- and between-block analyses. The pooled estimate is obtained using REML (restricted maximum likelihood).

PROGRAM

```
 2   JOB        'Example I'
 3
 4   OPEN       'XI'; 2
 5
 6   "Data:
-7   C    2.51   HM   2.15
-8   ...
-9   VM   2.28   LM   2.56 : "
10
11   UNITS      [30]
12   VARIATE    LDW
13   TEXT       NMDM; !T(C,'HM','AM','TM','VM','LM')
14   FACTOR     [LABELS=NMDM] MEDIUM
15   READ       [CHANNEL=2]  MEDIUM,LDW; FREP=Labels

     Identifier   Minimum     Mean   Maximum    Values   Missing
           LDW      1.700    2.288     2.650        30         0
16
17   FACTOR     [LEVELS=15] EMBRYO; !(2(1...15))
18   BLOCK      EMBRYO
19   TREATMENT  MEDIUM
20
21   "ANOVA produces the between- and within-block estimates required. To display
-22  the effects in the two strata, which are not output by default, the PRINT
-23  requirements are given explicitly:"
24
25   ANOVA      [PRINT=AOVTABLE,EFFECTS,INFORMATION,MEANS] LDW
```

```
***** Analysis of variance *****

Variate: LDW

Source of variation       d.f.      s.s.        m.s.      v.r.

EMBRYO stratum
MEDIUM                       5    0.613850    0.122770     7.95
Residual                     9    0.139030    0.015448     2.35

EMBRYO.*Units* stratum
MEDIUM                       5    0.446200    0.089240    13.56
Residual                    10    0.065800    0.006580

Total                       29    1.264880
```

```
***** Information summary *****

Model term                e.f.   non-orthogonal terms

EMBRYO stratum
  MEDIUM                  0.400

EMBRYO.*Units* stratum
  MEDIUM                  0.600   EMBRYO
```

```
* MESSAGE: the following units have large residuals.

EMBRYO 8          -0.149   s.e.  0.068


***** Tables of effects *****

Variate: LDW

***** EMBRYO stratum *****

MEDIUM effects    e.s.e. 0.0879  rep. 5

   MEDIUM        C       HM       AM       TM       VM       LM
             0.272   -0.280   -0.123   -0.060   -0.148    0.337

***** EMBRYO.*Units* stratum *****

MEDIUM effects   e.s.e. 0.0468  rep. 5

   MEDIUM        C       HM       AM       TM       VM       LM
             0.262    0.043   -0.092   -0.087   -0.228    0.102


***** Tables of means *****

Variate: LDW

Grand mean  2.288

   MEDIUM        C       HM       AM       TM       VM       LM
             2.550    2.331    2.196    2.201    2.060    2.390


*** Standard errors of differences of means ***

Table            MEDIUM
rep.                  5
s.e.d.           0.0662

   26  "REML (Residual Maximum Likelihood) calculates the combined estimates:"
   27
   28  VCOMPONENTS  [FIXED=MEDIUM] EMBRYO
   29  REML         [PRINT=EFFECTS] LDW


*** Table of effects for MEDIUM ***


      MEDIUM          C        HM        AM        TM        VM        LM
                 0.0000   -0.2647   -0.3591   -0.3461   -0.4803   -0.1288

Standard error of differences:    0.06368


   30  "The effects may be extracted and the estimates for each amino acid obtained:"
   31
   32  AKEEP      [STRATUM=EMBRYO] TERMS=MEDIUM; EFFECTS=BT
   33  AKEEP      TERMS=MEDIUM; EFFECTS=WT
   34  VKEEP      MEDIUM; EFFECTS=PT
   35  SCALAR     CW,CB,CP
   36  VARIATE    [NVAL=5] W,B,P,Within,Between,Pooled
   37  EQUATE     WT,BT,PT; !P(CW,W),!P(CB,B),!P(CP,P)
   38  CALCULATE  Within,Between,Pooled = CW,CB,CP-W,B,P
   39  PRINT      [ORIENT=ACROSS] !T(His,Arg,Thr,Val,Lys),Within,Between,Pooled;    \
   40                             FIELD=10; DECIMALS=4          "Table I.2 App. Stat."
```

	His	Arg	Thr	Val	Lys
Within	0.2183	0.3533	0.3483	0.4900	0.1600
Between	0.5525	0.3950	0.3325	0.4200	-0.0650
Pooled	0.2647	0.3591	0.3461	0.4803	0.1288

```
 41   "The pooled estimates differ from those printed in App.Stat. because of the
-42   different method of estimation used by REML. The weighted estimates of App.
-43   Stat. can be obtained by Genstat; the details are not given here."
 44
 45   STOP
```

******** End of Example I. Maximum of 14828 data units used at line 38 (35162 left)

SUGGESTED FURTHER WORK

1. Reproduce the analysis by working from first principles as in *App. Stat.*
2. Investigate the effect of the large residual (embryo 8).
3. How would you examine for homogeneity of variance?

Example J

Factorial experiment on cycles to failure of worsted yarn

DESCRIPTION OF DATA

In an unpublished report to the Technical Committee, International Wool Textile Organization, A. Barella and A. Sust gave the data in the first four columns of Table J.1, concerning the number of cycles to failure of lengths of worsted yarn under cycles of repeated loading. The three factors which varied over levels specified in coded form in the first three columns, are:

x_1, length of test specimen   (250, 300, 350 mm);
x_2, amplitude of loading cycle   (8, 9, 10 mm);
x_3, load   (40, 45, 50 g).

Table J.1 Cycles to failure, transformed values, fitted values and residuals

			Cycles	Log cycles		
x_1	x_2	x_3	obs	obs	fitted	resid
−1	−1	−1	674	6.51	6.52	−0.01
−1	−1	0	370	5.91	6.11	−0.20
−1	−1	1	292	5.68	5.74	−0.06
−1	0	−1	338	5.82	5.85	−0.03
−1	0	0	266	5.58	5.44	0.14
−1	0	1	210	5.35	5.07	0.28
−1	1	−1	170	5.14	5.26	−0.12
−1	1	0	118	4.77	4.84	−0.07
−1	1	1	90	4.50	4.48	0.02

Table J.1 (*cont.*)

| x_1 | x_2 | x_3 | Cycles | Log cycles | | |
			obs	obs	fitted	resid.
0	−1	−1	1414	7.25	7.42	−0.17
0	−1	0	1198	7.09	7.01	0.08
0	−1	1	634	6.45	6.64	−0.19
0	0	−1	1022	6.93	6.76	0.17
0	0	0	620	6.43	6.34	0.09
0	0	1	438	6.08	5.97	0.11
0	1	−1	442	6.09	6.16	−0.07
0	1	0	332	5.81	5.75	0.06
0	1	1	220	5.39	5.38	0.01
1	−1	−1	3636	8.20	8.18	0.02
1	−1	0	3184	8.07	7.77	0.30
1	−1	1	2000	7.60	7.40	0.20
1	0	−1	1568	7.36	7.52	−0.16
1	0	0	1070	6.98	7.11	−0.13
1	0	1	566	6.34	6.74	−0.40
1	1	−1	1140	7.04	6.92	0.12
1	1	0	884	6.78	6.51	0.27
1	1	1	360	5.89	6.14	−0.25

x_1, length of test specimen, x_2, amplitude of loading cycle, x_3, load.

THE ANALYSIS

The problem would be computationally straightforward if analysed as a 3^3 experiment with equal spacing of the levels of each factor, but taking log Y as dependent on $\log x_1$, $\log x_2$ and $\log x_3$ as in *App. Stat.*, destroys the advantages of the equal spacing on x_1, x_2 and x_3.

Only one column of Table J.1, the observed cycles to failure, needs to be input; the levels of the three factors and log cycles are generated by the program.

The full three-way table of data is displayed. Two-way and one-way means for preliminary inspection, as in Table J.2, *App. Stat.*, are obtained by the use of CALCULATE.

Orthogonal polynomials for the unequally spaced levels of the 3^3 design are generated by POLND; the arguments of POLND control which interactions with components of other factors are required. The analysis of variance, Table J.3, *App. Stat.,* is readily obtained by ANOVA.

A regression analysis based on the values of the orthogonal polynomials is also carried out.

PROGRAM

```
 2  JOB      'Example J'
 3
 4  OPEN     'XJ'; 2
 5
 6  "Data:
-7    674  370  292  338  266  210  170  118   90
-8   1414 1198  634 1022  620  438  442  332  220
-9   3636 3184 2000 1568 1070  566 1140  884  360 :
-10
-11  Only the observed numbers of cycles need to be read in; the systematic
-12  values of the factors can be generated. Numerical levels for the factors
-13  are given explicitly as variates; in the initial examination of the
-14  data these will be used to label rows and columns of tables, and later
-15  transformed for the analysis of variance."
 16
 17  UNITS     [27]
 18  VARIATE   LLTS,LALC,LLOAD; !(250,300,350),!(8,9,10),!(40,45,50)
 19  FACTOR    [LEVELS=LLTS]   LTS
 20  &         [LEVELS=LALC]   ALC
 21  &         [LEVELS=LLOAD]  LOAD
 22  READ      [CHANNEL=2]  CYCLES
```

Identifier	Minimum	Mean	Maximum	Values	Missing	
CYCLES	90.0	861.3	3636.0	27	0	Skew

```
 23
 24  GENERATE  LTS,ALC,LOAD
 25
 26  "We shall use the natural logarithms of the numbers of cycles throughout, so:"
 27
 28  CALCULATE LCYCLES = LOG(CYCLES)
 29
 30  "The data are tabulated by specifying a classification set (LTS, ALC and LOAD),
-31  the variate to be tabulated, and names of tables structures. By default, tables
-32  of totals will be formed, but other statistics - Mean, Variance, Minimum, or
-33  Maximum - may be requested."
 34
 35  TABULATE  [CLASS=LTS,ALC,LOAD] LCYCLES; MEANS=TCYCLES
 36
 37  "The table TCYCLES will be formed without margins; to set up space for the
-38  margins and fill them with means (or other statistics) over the factors the
-39  MARGIN directive is used:"
 40
 41  MARGIN    TCYCLES; METHOD=MEANS
 42
 43  "(Alternatively, the table TCYCLE can be declared with margins before the
-44  TABULATE instruction.)"
 45
 46  PRINT     TCYCLES; 10; 2
```

		TCYCLES			
LTS	LOAD	40.00	45.00	50.00	Margin
	ALC				
250.00	8.00	6.51	5.91	5.68	6.03
	9.00	5.82	5.58	5.35	5.58
	10.00	5.14	4.77	4.50	4.80
	Margin	5.82	5.42	5.17	5.47

300.00	8.00	7.25	7.09	6.45	6.93
	9.00	6.93	6.43	6.08	6.48
	10.00	6.09	5.81	5.39	5.76
	Margin	6.76	6.44	5.98	6.39
350.00	8.00	8.20	8.07	7.60	7.96
	9.00	7.36	6.98	6.34	6.89
	10.00	7.04	6.78	5.89	6.57
	Margin	7.53	7.28	6.61	7.14
Margin	8.00	7.32	7.02	6.58	6.97
	9.00	6.70	6.33	5.92	6.32
	10.00	6.09	5.79	5.26	5.71
	Margin	6.70	6.38	5.92	6.33

```
 47   "This displays all the information required. However, the two-way tables
-48   shown in App. Stat. can easily be obtained. First declare three tables with
-49   appropriate classifying factors."
 50
 51   TABLE     [CLASS=LOAD,ALC; MARGIN=YES] T1
 52   &         [CLASS=LOAD,LTS; MARGIN=YES] T2
 53   &         [CLASS=ALC,LTS; MARGIN=YES]  T3
 54
 55   "If we now CALCULATE T1=TCYCLES, the classifying sets of the two tables
-56   will be compared and since one factor - LTS - is not in the classifying
-57   set of T1, the marginal values over that factor will be taken from TCYCLES"
 58
 59   CALCULATE T1,T2,T3 - TCYCLES
 60   PRINT     T1,T2,T3; 10; 2                                    "Table J.2 App. Stat"
```

T1

ALC	8.00	9.00	10.00	Margin
LOAD				
40.00	7.32	6.70	6.09	6.70
45.00	7.02	6.33	5.79	6.38
50.00	6.58	5.92	5.26	5.92
Margin	6.97	6.32	5.71	6.33

T2

LTS	250.00	300.00	350.00	Margin
LOAD				
40.00	5.82	6.76	7.53	6.70
45.00	5.42	6.44	7.28	6.38
50.00	5.17	5.98	6.61	5.92
Margin	5.47	6.39	7.14	6.33

T3

LTS	250.00	300.00	350.00	Margin
ALC				
8.00	6.03	6.93	7.96	6.97
9.00	5.58	6.48	6.89	6.32
10.00	4.80	5.76	6.57	5.71
Margin	5.47	6.39	7.14	6.33

```
 61   "From this point, we will use the (natural) logarithms of the factor levels."
 62
 63   CALCULATE LLTS,LALC,LLOAD = LOG(LLTS,LALC,LLOAD)
 64
 65   "We now illustrate the ease with which different models may be fitted by
-66   Genstat's Analysis of Variance section. We fit the full factorial model,
-67   assigning the three factor interaction to error using the FACTORIAL option.
-68   The output is restricted to the AOV table only by the PRINT option."
 69
 70   TREATMENT LTS*ALC*LOAD
```

```
  71   ANOVA      [PRIN=AOV; FACTORIAL=2] LCYCLES
```

***** Analysis of variance *****

Variate: LCYCLES

Source of variation	d.f.	s.s.	m.s.	v.r.
LTS	2	12.51560	6.25780	301.37
ALC	2	7.17023	3.58511	172.65
LOAD	2	2.80023	1.40011	67.43
LTS.ALC	4	0.40118	0.10030	4.83
LTS.LOAD	4	0.13577	0.03394	1.63
ALC.LOAD	4	0.01457	0.00364	0.18
Residual	8	0.16612	0.02076	
Total	26	23.20369		

```
  72   "To fit linear and quadratic terms for the factors, a function POL or POLND
 -73   can be given in the treatments formula; POLND is appropriate here, and in
 -74   the following ANOVA statement the option CONTRASTS=2 restricts the contrasts
 -75   fitted to linear, linear x linear, and quadratic."
  76
  77   TREATMENT POLND(LTS; 2)*POLND(ALC; 2)*POLND(LOAD; 2)
  78   ANOVA      [PRINT=AOV; FACTORIAL=2; CONTRASTS=2; DEVIATIONS=0] LCYCLES      \
  79                                                      "Table J.3 App. Stat."
```

***** Analysis of variance *****

Variate: LCYCLES

Source of variation	d.f.	s.s.	m.s.	v.r.
LTS	2	12.51560	6.25780	166.04
Lin	1	12.51407	12.51407	332.04
Quad	1	0.00153	0.00153	0.04
ALC	2	7.17023	3.58511	95.13
Lin	1	7.16951	7.16951	190.23
Quad	1	0.00071	0.00071	0.02
LOAD	2	2.80023	1.40011	37.15
Lin	1	2.75243	2.75243	73.03
Quad	1	0.04780	0.04780	1.27
LTS.ALC	1	0.02025	0.02025	0.54
Lin.Lin	1	0.02025	0.02025	0.54
Assigned to error	3	0.38093	0.12698	3.37
LTS.LOAD	1	0.05169	0.05169	1.37
Lin.Lin	1	0.05169	0.05169	1.37
Assigned to error	3	0.08408	0.02803	0.74
ALC.LOAD	1	0.00500	0.00500	0.13
Lin.Lin	1	0.00500	0.00500	0.13
Assigned to error	3	0.00956	0.00319	0.08
Residual	17	0.64069	0.03769	
Total	26	23.20369		

```
  80   "To obtain the same analysis by regression, the factor values should be
 -81   replaced by othogonal polynomials; the linear by linear interactions have
 -82   to be calculated separately."
  83
  84   VARIATE    VALC,VLTS,VLOAD
  85   CALCULATE  VALC,VLTS,VLOAD = ALC,LTS,LOAD
  86   ORTHPOL    [MAX=2] VALC,VLTS,VLOAD; PALC,PLTS,PLOAD
  87   CALCULATE  I12,I23,I31 = PLTS[1],PALC[1],PLOAD[1] * PALC[1],PLOAD[1],PLTS[1]
  88   MODEL      LCYCLES
  89   TERMS      LCYCLES+PLTS[]+PALC[]+PLOAD[]+I12+I23+I31
```

```
90  "First, fit the linear terms"
91  FIT        PLTS[1]+PALC[1]+PLOAD[1]
```

***** Regression Analysis *****

Response variate: LCYCLES
     Fitted terms: Constant + PLTS[1] + PALC[1] + PLOAD[1]

*** Summary of analysis ***

	d.f.	s.s.	m.s.	v.r.
Regression	3	22.4360	7.47867	224.06
Residual	23	0.7677	0.03338	
Total	26	23.2037	0.89245	
Change	-3	-22.4360	7.47867	224.06

Percentage variance accounted for 96.3

* MESSAGE: The following units have large standardized residuals:
            24   -2.36

*** Estimates of regression coefficients ***

	estimate	s.e.	t
Constant	6.3347	0.0352	180.17
PLTS[1]	4.950	0.256	19.36
PALC[1]	-5.654	0.386	-14.66
PLOAD[1]	-3.503	0.386	-9.08

```
 92  "Quadratic terms and the interactions can be added by:
-93  ADD       PLTS[2]+PALC[2]+PLOAD[2]+I12+I23+I31   The output is not shown here.
-94  Residuals can be plotted and examined, as in Examples G and P."
 95
 96  STOP
```

******** End of Example J. Maximum of 16576 data units used at line 86 (33414 left)

SUGGESTED FURTHER WORK

Compare the fit of $\log Y$ on x_1, x_2, x_3 with that of $\log Y$ on $\log x_1$, $\log x_2$, $\log x_3$ (using linear terms only).

Example K

Factorial experiment on diets for chickens

DESCRIPTION OF DATA

An experiment comparing 12 methods of feeding chickens (Duckworth and Carpenter; see John and Quenouille, 1977) was done independently in two replicates arranged in different houses. The treatments, forming a $3 \times 2 \times 2$ factorial, were 'form of protein', 'level of protein', 'level of fish solubles'. The data are given in Table K.1.

Table K.1 Total weights of 16 six-week-old chicks (g)

Protein	Level of protein	Level of fish solubles	House	
			I	II
Groundnut	0	0	6559	6292
		1	7075	6779
	1	0	6564	6622
		1	7528	6856
	2	0	6738	6444
		1	7333	6361
Soyabean	0	0	7094	7053
		1	8005	7657
	1	0	6943	6249
		1	7359	7292
	2	0	6748	6422
		1	6764	6560

THE ANALYSIS

The design is a balanced factorial experiment with 2 proteins $\times$ 3 levels of protein $\times$ 2 fish solubles, replicated in 2 houses.

Tables of two-way and one-way marginal means for preliminary inspection, Table K.2, *App. Stat.*, are produced using TABULATE and MARGIN.

An analysis of variance table is computed separately for the $2 \times 3 \times 2$ factorial experiment within each house, Table K.3(a), *App. Stat.* By declaring 'house' to be a block effect the pooled analysis of variance, Table K.3(b), *App. Stat.*, is obtained.

High-quality graphics illustrate the observed interaction effects; see Fig. K.1 after the Genstat program.

PROGRAM

```
 2   JOB        'Example K'
 3
 4   OPEN       'XK'; 2
 5
 6   "Data:
-7   6559 6292
-8   ...
-9   6764 6560 : "
10
11   UNITS      [24]
12   READ       [CHANNEL=2] WEIGHT
```

```
     Identifier   Minimum     Mean   Maximum    Values   Missing
        WEIGHT       6249      6887      8005        24         0
```

```
13
14   TEXT       NP; !T(Groundnut,Soyabean)
15   VARIATE    LS; !(0,1)
16   &          LP; !(0,1,2)
17   FACTOR     [LABELS=NP] PROTEIN
18   FACTOR     [LEVELS=LP] PLEVEL
19   FACTOR     [LEVELS=LS] SLEVEL
20   FACTOR     [LEVELS=2] HOUSE
21
22   "The data are  available in standard order, and the appropriate values
-23   for the factors are generated:"
24
25   GENERATE   PROTEIN,PLEVEL,SLEVEL,HOUSE
26
27   "Most of the two-way means given in Table K.2 App. Stat. will appear during
-28   a subsequent analysis of variance, but the tables are easily formed by
-29   TABULATE and displayed in the form shown in App. Stat."
30
31   TABLE      [CLASS=PROTEIN,HOUSE]   MPH
32   &          [CLASS=PLEVEL,HOUSE]    MTH
33   &          [CLASS=SLEVEL,HOUSE]    MSH
34   &          [CLASS=PROTEIN,PLEVEL]  MPT
35   &          [CLASS=PROTEIN,SLEVEL]  MPS
36   &          [CLASS=SLEVEL,PLEVEL]   MST
37   TABULATE   WEIGHT; MEANS = MPH
38   &          WEIGHT; MEANS = MTH
39   &          WEIGHT; MEANS = MSH
40   &          WEIGHT; MEANS = MPT
41   &          WEIGHT; MEANS = MPS
42   &          WEIGHT; MEANS = MST
43   MARGIN     MPT,MPS,MSH,MST; METHOD=MEANS
44   PRINT      MPH,MTH,MSH,MPT,MPS,MST; FIELD=9; DECIMALS=0      "Table K.2 App. Stat"
```

	MPH	
HOUSE	1	2
PROTEIN		
Groundnut	6966	6559
Soyabean	7152	6872

```
                   MTH
        HOUSE       1       2
        PLEVEL
         0.00     7183    6945
         1.00     7099    6755
         2.00     6896    6447


                   MSH
        HOUSE       1       2     Margin
        SLEVEL
         0.00     6774    6514    6644
         1.00     7344    6918    7131

        Margin    7059    6716    6887


                   MPT
        PLEVEL     0.00    1.00    2.00    Margin
        PROTEIN
      Groundnut    6676    6893    6719    6763
       Soyabean    7452    6961    6624    7012

        Margin    7064    6927    6671    6887


                   MPS
        SLEVEL     0.00    1.00   Margin
        PROTEIN
      Groundnut    6537    6989    6763
       Soyabean    6752    7273    7012

        Margin    6644    7131    6887


                   MST
        PLEVEL     0.00    1.00    2.00    Margin
        SLEVEL
         0.00     6750    6595    6588    6644
         1.00     7379    7259    6755    7131

        Margin    7064    6927    6671    6887
```

```
  45   "The mean squares in the analysis of variance in Table K.3 App.Stat. are scaled
 -46   down by 10**6; the same effect is achieved by scaling WEIGHT by 10**3."
  47
  48   CALCULATE WEIGHT = WEIGHT / 1000
  49
  50   "The obvious analysis, pooling over houses, is obtained by treating HOUSE
 -51   as a random effect, and the other factors as treatments in a full 2x3x2
 -52   factorial design."
  53
  54   BLOCK       HOUSE
  55   TREATMENT   PROTEIN*PLEVEL*SLEVEL
  56   ANOVA       WEIGHT                                    "Table K3(b) App. Stat."
```

***** Analysis of variance *****

Variate: WEIGHT

Source of variation	d.f.	s.s.	m.s.	v.r.
HOUSE stratum	1	0.70830	0.70830	15.82
HOUSE.*Units* stratum				
PROTEIN	1	0.37375	0.37375	8.35
PLEVEL	2	0.63628	0.31814	7.10
SLEVEL	1	1.42155	1.42155	31.74
PROTEIN.PLEVEL	2	0.85816	0.42908	9.58
PROTEIN.SLEVEL	1	0.00718	0.00718	0.16
PLEVEL.SLEVEL	2	0.30889	0.15444	3.45
PROTEIN.PLEVEL.SLEVEL	2	0.05013	0.02506	0.56
Residual	11	0.49264	0.04479	
Total	23	4.85687		

```
* MESSAGE: the following units have large residuals.

HOUSE 1     *units* 6           0.314   s.e. 0.143
HOUSE 2     *units* 6          -0.314   s.e. 0.143


***** Tables of means *****

Variate: WEIGHT

Grand mean  6.887

    PROTEIN Groundnut  Soyabean
                6.763     7.012

    PLEVEL      0.00      1.00      2.00
                7.064     6.927     6.671

    SLEVEL      0.00      1.00
                6.644     7.131

    PROTEIN    PLEVEL     0.00      1.00      2.00
  Groundnut               6.676     6.892     6.719
  Soyabean                7.452     6.961     6.623

    PROTEIN    SLEVEL     0.00      1.00
  Groundnut               6.536     6.989
  Soyabean                6.752     7.273

    PLEVEL     SLEVEL     0.00      1.00
    0.00                  6.750     7.379
    1.00                  6.595     7.259
    2.00                  6.588     6.754

               PLEVEL     0.00                1.00                2.00
    PROTEIN    SLEVEL     0.00      1.00      0.00      1.00      0.00      1.00
  Groundnut               6.426     6.927     6.593     7.192     6.591     6.847
  Soyabean                7.074     7.831     6.596     7.325     6.585     6.662


*** Standard errors of differences of means ***

Table               PROTEIN     PLEVEL      SLEVEL      PROTEIN
                                                        PLEVEL
rep.                   12          8           12          4
s.e.d.               0.0864      0.1058      0.0864      0.1496

Table               PROTEIN     PLEVEL      PROTEIN
                    SLEVEL      SLEVEL      PLEVEL
                                            SLEVEL
rep.                   6           4           2
s.e.d.               0.1222      0.1496      0.2116


  57   "To analyse each house separately, we cancel the use of HOUSE as a random
 -58   factor and restrict the analysis to the values in House 1 and House 2
 -59   separately. The sums of squares needed are extracted using AKEEP; the
 -60   first use of this command also gets the degrees of freedom. Each value
 -61   has to be placed in a scalar (automatically declared by AKEEP); these
 -62   will be later put together as variates of length 7 for convenience"
  63
  64   BLOCK
  65   AKEEP      PROTEIN*PLEVEL*SLEVEL; SS=SSM[1...7]; DF=DF[1...7]
  66   RESTRICT   WEIGHT; HOUSE.EQ.1
  67   ANOVA      [PRIN=AOV]  WEIGHT
```

***** Analysis of variance *****

Variate: WEIGHT

Source of variation	d.f.	s.s.	m.s.	v.r.
PROTEIN	1	0.103788	0.103788	
PLEVEL	2	0.174595	0.087298	
SLEVEL	1	0.973561	0.973561	
PROTEIN.PLEVEL	2	0.521914	0.260957	
PROTEIN.SLEVEL	1	0.044652	0.044652	
PLEVEL.SLEVEL	2	0.104952	0.052476	
PROTEIN.PLEVEL.SLEVEL	2	0.153241	0.076620	
Total	11	2.076702		

```
   68  AKEEP     PROTEIN*PLEVEL*SLEVEL; SS=SS1[1...7]
   69  RESTRICT  WEIGHT; HOUSE.EQ.2
   70  ANOVA     [PRIN=AOV]  WEIGHT
```

***** Analysis of variance *****

Variate: WEIGHT

Source of variation	d.f.	s.s.	m.s.	v.r.
PROTEIN	1	0.294220	0.294220	
PLEVEL	2	0.506209	0.253104	
SLEVEL	1	0.489244	0.489244	
PROTEIN.PLEVEL	2	0.386185	0.193092	
PROTEIN.SLEVEL	1	0.109634	0.109634	
PLEVEL.SLEVEL	2	0.216764	0.108382	
PROTEIN.PLEVEL.SLEVEL	2	0.069619	0.034809	
Total	11	2.071875		

```
   71  AKEEP     PROTEIN*PLEVEL*SLEVEL; SS=SS2[1...7]
   72  VARIATE   df,Main,House1,House2; !(DF[]),!(SSM[]),!(SS1[]),!(SS2[])
   73  CALCULATE Inter = House1+House2-Main
   74  &         House1,House2,Main,Inter = House1,House2,Main,Inter / df
   75  PRINT                                                              \
   76       !T('P      ','Lp     ','Lf     ','P.Lp  ','P.Lf ','Lp.Lf ','P.Lp.Lf'), \
   77       df,House1,House2,Main,Inter; FIELD=9; DECIMALS=2(0),4(4)      \
   78                                                 "Table K.3(a) App. Stat."
```

	df	House1	House2	Main	Inter
P	1	0.1038	0.2942	0.3738	0.0243
Lp	2	0.0873	0.2531	0.3181	0.0223
Lf	1	0.9736	0.4892	1.4216	0.0413
P.Lp	2	0.2610	0.1931	0.4291	0.0250
P.Lf	1	0.0447	0.1096	0.0072	0.1471
Lp.Lf	2	0.0525	0.1084	0.1544	0.0064
P.Lp.Lf	2	0.0766	0.0348	0.0251	0.0864

```
   79  "To reproduce Figure K.1 App. Stat., we use TABULATE to form a three-way table
  -80  of means over HOUSE, and extract the four rows of this table as variates. WEIGHT
  -81  must be de-restricted, and rescaled to match Table K.1 App. Stat."
   82
   83  RESTRICT  WEIGHT
   84  CALCULATE WEIGHT=10*WEIGHT
   85  TABULATE  [CLASS=PROTEIN,SLEVEL,PLEVEL] WEIGHT; MEANS=TM
   86  VARIATE   [NVALUES=3] G0,G1,S0,S1
   87  EQUATE    TM; !P(G0,G1,S0,S1)
   88  OPEN      'K.PLT'; 6; Graphics
   89  DEVICE    6
   90  AXES      WINDOW=1; XTITLE='Level of protein'; YTITLE='Weight (gr/100)'; \
   91            YLOWER=60; YUPPER=80; XLOWER=0; XUPPER=2; XINTEGER=Y; YINTEGER=Y; \
   92            XMARKS=!(0,1,2); YMARKS=!(65,70...80)
   93  PEN       1,2,3,4; LINESTYLE=1,2,3,4; METHOD=LINE
   94  DGRAPH    G0,G1,S0,S1; LP; PEN=1,2,3,4;                              \
   95            DESCRIPTION=!T('Groundnut; Level of fish solubles, 0'),    \
   96                        !T('Groundnut; Level of fish solubles, 1'),    \
   97                        !T('Soyabean;  Level of fish solubles, 0'),    \
```

```
98                    !T('Soyabean;  Level of fish solubles, 1')
99   STOP
```

```
******** End of Example K.  Maximum of 14358 data units used at line 71 (35632 left)
```

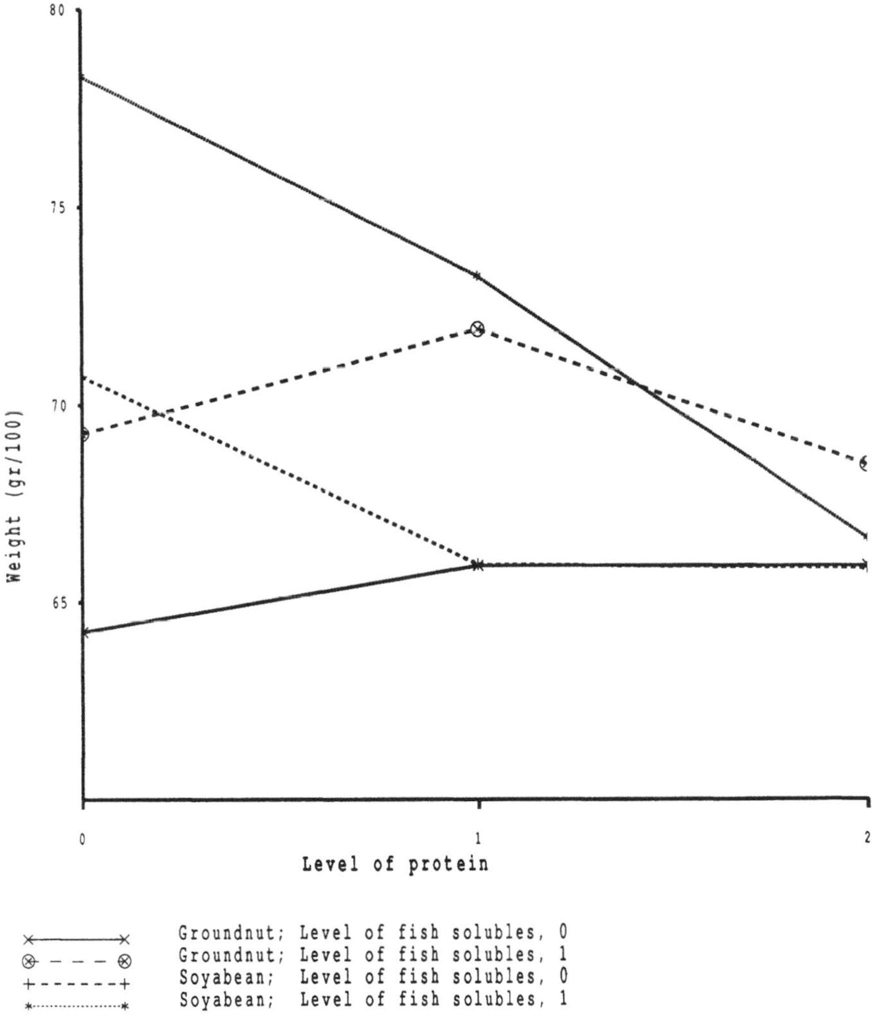

```
×————×      Groundnut;  Level of fish solubles, 0
⊗ – – – ⊗    Groundnut;  Level of fish solubles, 1
+ – – – – – +  Soyabean;   Level of fish solubles, 0
*·············*  Soyabean;   Level of fish solubles, 1
```

Figure K.1

SUGGESTED FURTHER WORK

1. Repeat the analysis using log weight. Comment.
2. Carry out an analysis using multiple regression with protein and fish-soluble levels each coded as −1, 1 and the three levels of protein as −1, 0, 1. Compare the results.

Example L

Binary preference data for detergent use

DESCRIPTION OF DATA

Table L.1 (Ries and Smith, 1963) compares two detergents, a new product X and a standard product M. Each individual expresses a preference between X and M. In the table, Y_j is the number of individuals out of n_j in 'cell' j who prefer X, the remaining $n_j - Y_j$ preferring M. The individuals are classified by three factors, water softness at three levels, temperature at two levels, and a factor whose two levels correspond to previous experience and no previous experience with M. The object is to study how preferences for X vary.

Table L.1 Number Y_j of preferences for brand X out of n_j individuals

Water softness		M previous non-user		M previous user	
		Temperature		Temperature	
		Low	High	Low	High
Hard	Y_j	68	42	37	24
	n_j	110	72	89	67
Medium	Y_j	66	33	47	23
	n_j	116	56	102	70
Soft	Y_j	63	29	57	19
	n_j	116	56	106	48

THE ANALYSIS

The discussion in *App. Stat.* points out the range in proportion of preferences for brand $X(Y_j/n_j)$ is such that an analysis of the untransformed proportions is reasonable and has the advantage of direct interpretation. If, however, the results are to be compared with another data set having proportions at a higher or lower level, then a logistic analysis is likely to be preferable. Both methods of analysis are illustrated here.

The observed proportions and their estimated variances, $\{Y_j(n_j - Y_j)\}/\{n_j^2(n_j - 1)\}$, are computed. Two-way and one-way means are obtained using the MARGIN command.

PROGRAM

```
 2   JOB        'Example L'
 3
 4   OPEN       'XL'; 2
 5
 6   "Data:
-7    68 110   42  72   37  89   24  67
-8    66 116   33  56   47 102   23  70
-9    63 116   29  56   57 106   19  48 : "
10
11   UNITS      [12]
12   TEXT       NT; !T(Low,High)
13    &         NU; !T('Non-user','User')
14    &         NW; !T(Hard,Medium,Soft)
15   FACTOR     [LABELS=NT]  TEMP
16    &         [LABELS=NU]  Previous
17    &         [LABELS=NW]  WATER
18   TABLE      [CLASS=WATER,Previous,TEMP] Y,N,PRP,VAR
19   READ       [CHANNEL=2]  Y,N
```

Identifier	Minimum	Mean	Maximum	Values	Missing
Y	19.00	42.33	68.00	12	0
N	48.00	84.00	116.00	12	0

```
20
21   CALCULATE  PRP = Y/N
22    &         VAR = Y*(N-Y)/(N*N*(N-1))
23    &         M=MEAN(VAR)
24
25   "We shall need these values later; this is a convenient moment to form them."
26
27   EQUATE     Y,N,PRP; OBS,NOS,RATIOS
28
29   "Options of print are used to suppress printing of the structure names
-30  (IPRINT=*) and determine the position of the structure list in the list
-31  of factors forming the classification set of the tables (INTERLEAVE)."
32
33   PRINT      [IPRINT=*; INTERLEAVE=2] PRP,VAR; FIELD=10; DEC=4,5            \
34                                                   "Table L.2 App. Stat."
```

	Previous	Non-user		User	
	TEMP	Low	High	Low	High
WATER					
Hard		0.6182	0.5833	0.4157	0.3582
		0.00217	0.00342	0.00276	0.00348
Medium		0.5690	0.5893	0.4608	0.3286
		0.00213	0.00440	0.00246	0.00320
Soft		0.5431	0.5179	0.5377	0.3958
		0.00216	0.00454	0.00237	0.00509

```
35    "We compute two-way and one-way means:"
36
37    TABLE      [CLASS=TEMP,Previous;MARGIN=YES]  TU
38    &          [CLASS=WATER,Previous;MARGIN=YES]  WU
39    &          [CLASS=WATER,TEMP;MARGIN=YES]     WT
40    MARGIN     PRP; METHOD=MEANS
41    CALCULATE  WT,WU,TU = PRP
42    PRINT      WT,WU,TU; FIELD=10; DEC=3                      "Table L.3 App. Stat."
```

	WT		
TEMP	Low	High	Margin
WATER			
Hard	0.517	0.471	0.494
Medium	0.515	0.459	0.487
Soft	0.540	0.457	0.499
Margin	0.524	0.462	0.493

	WU		
Previous	Non-user	User	Margin
WATER			
Hard	0.601	0.387	0.494
Medium	0.579	0.395	0.487
Soft	0.530	0.467	0.499
Margin	0.570	0.416	0.493

	TU		
Previous	Non-user	User	Margin
TEMP			
Low	0.577	0.471	0.524
High	0.563	0.361	0.462
Margin	0.570	0.416	0.493

```
43    "To use ANOVA, the factors must be given values; again, the arrangement of the
-44   data values is systematic, and the values can be generated:"
45
46    GENERATE  WATER,Previous,TEMP
47
48    "Use linear and quadratic orthogonal polynomials for water softness:"
49
50    TREATMENT POLND(WATER;2)*Previous*TEMP
51    ANOVA     [PRINT=AOV] RATIOS                              "Table L.4(a) App. Stat."
```

***** Analysis of variance *****

Variate: RATIOS

Source of variation	d.f.	s.s.	m.s.	v.r.
WATER	2	0.0002784	0.0001392	
Lin	1	0.0000455	0.0000455	
Quad	1	0.0002329	0.0002329	
Previous	1	0.0711269	0.0711269	
TEMP	1	0.0114955	0.0114955	
WATER.Previous	2	0.0126565	0.0063283	
Lin.Previous	1	0.0112638	0.0112638	
Quad.Previous	1	0.0013927	0.0013927	
WATER.TEMP	2	0.0007522	0.0003761	
Lin.TEMP	1	0.0006990	0.0006990	
Quad.TEMP	1	0.0000532	0.0000532	
Previous.TEMP	1	0.0070986	0.0070986	
WATER.Previous.TEMP	2	0.0022486	0.0011243	
Lin.Previous.TEMP	1	0.0011041	0.0011041	
Quad.Previous.TEMP	1	0.0011445	0.0011445	
Total	11	0.1056568		

```
52    "There are no entries in the variance ratio column because a full model has
-53   been fitted and no residual mean square is available. For the analysis, split
-54   for previous non-user and user, the  factor Previous should be removed from
-55   the treatment formula, and the analysis of variance repeated with RATIOS
-56   restricted to units with previous non-user and user respectively."
```

```
57
58   TREATMENT POLND(WATER; 2)*TEMP
59   RESTRICT  RATIOS; Previous.EQ.1
60   ANOVA     [PRINT=AOV] RATIOS
```

***** Analysis of variance *****

Variate: RATIOS

Source of variation	d.f.	s.s.	m.s.	v.r.
WATER	2	0.00518214	0.00259107	
Lin	1	0.00493890	0.00493890	
Quad	1	0.00024324	0.00024324	
TEMP	1	0.00026367	0.00026367	
WATER.TEMP	2	0.00086868	0.00043434	
Lin.TEMP	1	0.00002305	0.00002305	
Quad.TEMP	1	0.00084563	0.00084563	
Total	5	0.00631449		

```
61   "The corresponding analysis of variance for previous users can be obtained by:
-62  RESTRICT  RATIOS; Previous.EQ.2
-63  ANOVA     [PRINT=AOV] RATIOS
-64
-65  We now illustrate logistic regression. We need orthogonal polynomials:"
66
67   ORTHPOL   [MAX=2] WATER; PW
68   MODEL     [DISTRIBUTION=BINOMIAL; LINK=LOGIT] OBS; NBINOMIAL=NOS
69   TERMS     [FACTORIAL=3] (Previous*TEMP)*(PW[1]+PW[2])
70
71   "The full model can be fitted directly by
-72  FIT       [PRINT=ESTIMATES,ACCUMULATED] (TEMP*Previous)*(PW[1]+PW[2])
-73  but the deviances for the individual terms can only be obtained by a
-74  sequential process, adding in one term at a time. Printing of the
-75  analysis of variance and the estimates can be suppressed during the
-76  intermediate stages."
77
78   ADD       [PRINT=*]   Previous & TEMP & PW[1]
79   &         [PRINT=ACCUMULATED,ESTIMATES] PW[2]
```

***** Regression Analysis *****

*** Estimates of regression coefficients ***

	estimate	s.e.	t
Constant	0.382	0.100	3.81
Previous User	−0.567	0.128	−4.44
TEMP High	−0.257	0.133	−1.93
PW[1]	0.0106	0.0788	0.13
PW[2]	0.060	0.135	0.45

* MESSAGE: s.e.s are based on dispersion parameter with value 1

*** Accumulated analysis of deviance ***

Change	d.f.	deviance	mean deviance	deviance ratio
+ Previous	1	20.581	20.581	17.51
+ TEMP	1	3.800	3.800	3.23
+ PW[1]	1	0.017	0.017	0.01
+ PW[2]	1	0.199	0.199	0.17
Residual	7	8.228	1.175	
Total	11	32.826	2.984	

```
80   ADD   [PRINT=*]   Previous*TEMP & Previous*PW[1] & Previous*PW[2]
81   &                 TEMP*PW[1] & TEMP*PW[2] & Previous*TEMP*PW[1]
82   &     [PRINT=ESTIMATES,ACCUMULATED]       Previous*TEMP*PW[2]
```

***** Regression Analysis *****

*** Estimates of regression coefficients ***

	estimate	s.e.	t
Constant	0.311	0.110	2.83
Previous User	-0.426	0.161	-2.65
TEMP High	-0.054	0.186	-0.29
PW[1]	-0.154	0.135	-1.14
PW[2]	0.050	0.231	0.21
Previous User .TEMP High	-0.404	0.269	-1.50
PW[1].Previous User	0.400	0.198	2.02
PW[2].Previous User	0.013	0.338	0.04
PW[1].TEMP High	0.022	0.225	0.10
PW[2].TEMP High	-0.207	0.399	-0.52
PW[1].Previous User .TEMP High			
	-0.188	0.331	-0.57
PW[2].Previous User .TEMP High			
	0.356	0.568	0.63

* MESSAGE: s.e.s are based on dispersion parameter with value 1

*** Accumulated analysis of deviance ***

Change	d.f.	deviance	mean deviance	deviance ratio
+ Previous	1	20.5815	20.5815	52.37
+ TEMP	1	3.8002	3.8002	9.67
+ PW[1]	1	0.0168	0.0168	0.04
+ PW[2]	1	0.1992	0.1992	0.51
+ Previous.TEMP	1	2.7328	2.7328	6.95
+ PW[1].Previous	1	4.2973	4.2973	10.94
+ PW[2].Previous	1	0.2988	0.2988	0.76
+ PW[1].TEMP	1	0.1503	0.1503	0.38
+ PW[2].TEMP	1	0.0115	0.0115	0.03
+ PW[1].Previous.TEMP	1	0.3443	0.3443	0.88
Residual	1	0.3930	0.3930	
+ PW[2].Previous.TEMP	1	0.3930	0.3930	1.00
Total	11	32.8256	2.9841	

```
 83    "Separate analyses may be computed for previous non-users and users by using
-84    RESTRICT and deleting Previous from the FIT command."
 85
 86    STOP
```

******** End of Example L. Maximum of 17482 data units used at line 69 (31878 left)

SUGGESTED FURTHER WORK

1. Compute the logistic analysis separately for previous non-users and users of M.
2. Compare the results from the two methods of analysis.
3. How in general would one decide whether linear, or logistic, or some other model is preferable?

Example M

Fertilizer experiment on growth of cauliflowers

DESCRIPTION OF DATA

In an experiment on the effect of nitrogen and potassium upon the growth of cauliflowers, four levels of nitrogen and two levels of potassium were tested.

Nitrogen levels: 0, 60, 120, 180 units per acre (coded as 0, 1 2, 3)
Potassium levels: 200, 300 units per acre (coded as A, B).

The experiment was arranged in 4 blocks, each containing 4 plots and when harvested, the cauliflowers were graded according to size. Table M.1 shows the yield (number of cauliflowers) of different sizes: grade 12, for example, means that 12 cauliflowers fit into a standard size crate. The data were provided by Mr. J. C. Gower, Rothamsted Experimental Station.

THE ANALYSIS

The design of the experiment is a complete 2×4 factorial, confounded into two blocks I, II, with blocks III, IV forming a replicate of the design. The interaction of potassium with the quadratic component of nitrogen, i.e. $K \times N_Q$, defines the confounding.

In *App. Stat.* the analysis relies upon identifying the contrasts representing the 3 degrees of freedom between blocks. When using Genstat there is no need to do this; the ANOVA algorithm will indicate the departure from complete balance.

The response measure Y analysed in *App. Stat.* is the effective number of crates of marketable cauliflowers, given by

$$Y = \tfrac{1}{12}n_{12} + \tfrac{1}{16}n_{16} + \tfrac{1}{24}n_{24} + \tfrac{1}{30}n_{30}. \quad \text{(M.1, *App. Stat.*)}$$

An alternative response measure Z, the proportion of cauliflowers of grade 24 or better, given by

$$Z = (n_{12} + n_{16} + n_{24})/48 \qquad \text{(M.2, App. Stat.)}$$

is not analysed in *App. Stat.*

Instructions are given for analysing both the above response variables, Y by ordinary multiple regression and Z by logistic regression. Note, however, that the 'usual' formulae for standard errors will probably underestimate the true 'error' because of positive correlation between the qualities of cauliflowers in a plot.

Table M.1 Numbers of cauliflowers of each grade

Block	Treatment	12	16	24	30	Unmarketable
I	0A	—	1	21	24	2
	2B	1	6	24	13	4
	1B	—	4	28	12	4
	3A	1	10	26	9	1
II	3B	—	4	26	14	4
	1A	—	5	27	13	3
	0B	—	—	12	28	8
	2A	—	5	35	5	3
III	1B	—	1	22	22	3
	0A	—	1	8	33	3
	3A	—	6	22	17	2
	2B	—	3	27	14	4
IV	0B	—	—	8	30	10
	2A	—	7	16	22	3
	3B	—	2	31	11	4
	1A	—	—	13	26	9

The Grade columns are 12, 16, 24, 30.

PROGRAM

```
 2  JOB        'Example M'
 3
 4  OPEN       'XM'; 2
 5
 6  "Data:
-7  0 A    0  1 21 24  2
-8  ...
-9  1 A    0  0 13 26  9 :
-10
-11  The factors are initially declared with labels or levels that correspond
-12  to the data representation (A or B for Potassium, 0,1,2 or 3 for Nitrogen).
-13  Later the labels for K will be redefined to allow appropriate annotation
-14  in the output from ANOVA; and the levels of N will be multiplied by 60 to
-15  produce the correct numerical values."
```

```
16
17  UNITS      [16]
18  VARIATE    LK; !(200,300)
19    &        LN; !(0...3)
20  TEXT       NK; !T(A,B)
21    &        NN; !T('0','60','120','180')
22  FACTOR     [LEVELS=4]  BLOCK; !(4(1...4))
23    &        [LABELS=NK; LEVELS=LK]  K
24    &        [LABELS=NN; LEVELS=LN]  N
25  READ       [CHANNEL=2]  N,K,Y12,Y16,Y24,Y30,YUM; FREP=Levels,Labels,5(*)
```

Identifier	Minimum	Mean	Maximum	Values	Missing	
Y12	0.0000	0.1250	1.0000	16	0	Skew
Y16	0.000	3.438	10.000	16	0	
Y24	8.00	21.62	35.00	16	0	
Y30	5.00	18.31	33.00	16	0	
YUM	1.000	4.187	10.000	16	0	

```
26
27  CALCULATE Y = Y12/12+Y16/16+Y24/24+Y30/30
28  TEXT       NK; !T('200','300')
29  CALCULATE LN = 60*LN
30  PRINT      BLOCK,N,K,Y; FIELD=3(7),10; DECIMALS=3(0),3     "Table M.2 App. Stat."
```

BLOCK	N	K	Y
1	0	200	1.738
1	120	300	1.892
1	60	300	1.817
1	180	200	2.092
2	180	300	1.800
2	60	200	1.871
2	0	300	1.433
2	120	200	1.938
3	60	300	1.713
3	0	200	1.496
3	180	200	1.858
3	120	300	1.779
4	0	300	1.333
4	120	200	1.838
4	180	300	1.783
4	60	200	1.408

```
31   "The ANOVA algorithm determines the design from the factor values and the
-32  structure formulae. Any departure from complete balance will be indicated.
-33  If the design is too unbalanced ANOVA will abandon its task: the regression
-34  facilities should be used in that case."
35
36  BLOCK      BLOCK
37  TREATMENT  POLND(N;2)*K
38  ANOVA      [DEVIATIONS=0] Y
```

```
******** Warning (Code AN 17). Statement 1 on Line 38
Command: ANOVA [DEVIATIONS=0] Y
```

Partial confounding
N.K is partially confounded with BLOCK

***** Analysis of variance *****

Variate: Y

Source of variation	d.f.	s.s.	m.s.	v.r.
BLOCK stratum				
N.K	2	0.05992	0.02996	0.26
Lin.K	1	0.00000	0.00000	0.00
Quad.K	1	0.05992	0.05992	0.51
Assigned to error	1	0.00000	0.00000	0.00
Residual	1	0.11746	0.11746	9.51

```
BLOCK.*Units* stratum
N                          2    0.37537    0.18768    15.19
   Lin                     1    0.34289    0.34289    27.75
   Quad                    1    0.03248    0.03248     2.63
   Assigned to error       1    0.00180    0.00180     0.15
K                          1    0.02954    0.02954     2.39
N.K                        2    0.00004    0.00002     0.00
   Lin.K                   1    0.00004    0.00004     0.00
   Quad.K                  1    0.00000    0.00000     0.00
   Assigned to error       1    0.01689    0.01689     1.37
Residual                   7    0.08650    0.01236

Total                     15    0.66883
```

* MESSAGE: the following units have large residuals.

BLOCK 4   -0.204   s.e. 0.074

***** Tables of means *****

Variate: Y

Grand mean   1.737

```
        N        0       60      120      180
              1.495    1.716    1.847    1.888

        K      200      300
              1.780    1.694

        N      K      200      300
        0            1.536    1.454
       60            1.759    1.674
      120            1.891    1.804
      180            1.933    1.843
```


*** Standard errors of differences of means ***

```
Table              N           K           N
                                           K
                        (smoothed)
rep.               4           8           2
s.e.d.         0.0786      0.0556      0.1112
```


```
 39    "This, effectively, gives the results shown in Tables M.4 and M.5, App. Stat.
-40
-41    The partial confounding of the NK interaction is noted. A zero sums-of-squares
-42    for a term in a particular stratum implies that it is not estimable in that
-43    stratum: thus the interaction of N(Quad) and K is seen to be confounded with
-44    BLOCK and the s.s. for that stratum may be combined to give a total s.s. for
-45    BLOCK of 0.17738, with 3 d.f.
-46    The main effect of K, the linear and quadratic components of N, and the inter-
-47    action of K with the linear component of N are estimated in the lowest stratum.
-48
-49    This kind of confounding can also be handled by using pseudo factors.
-50    For details, see the Genstat Manual.
-51
-52    A logistic model can be fitted to the alternative derived response."
 53
 54    CALCULATE Z = (Y12+Y16+Y24)
 55    ORTHPOL   [MAX=2] N; PN
 56    MODEL     [DISTRIBUTION=BINOMIAL; LINK=LOGIT] Z; NBINOMIAL=!(16(48))
 57    TERMS     BLOCK+PN[1]*K+PN[2]
 58    ADD       [PRINT=*] BLOCK  & K  & PN[1]  & PN[2]
 59    &         [PRINT=ACCUMULATED,ESTIMATES] K*PN[1]
```

***** Regression Analysis *****

*** Estimates of regression coefficients ***

	estimate	s.e.	t
Constant	0.643	0.177	3.64
BLOCK 2	-0.197	0.223	-0.88
BLOCK 3	-0.754	0.220	-3.43
BLOCK 4	-1.084	0.224	-4.84
K 300	-0.083	0.156	-0.53
PN[1]	0.00904	0.00166	5.44
PN[2]	-0.0000738	0.0000218	-3.39
PN[1].K 300	0.00162	0.00240	0.67

* MESSAGE: s.e.s are based on dispersion parameter with value 1

*** Accumulated analysis of deviance ***

Change	d.f.	deviance	mean deviance	deviance ratio
+ BLOCK	3	27.491	9.164	3.56
+ K	1	0.135	0.135	0.05
+ PN[1]	1	72.525	72.525	28.15
+ PN[2]	1	11.425	11.425	4.43
+ PN[1].K	1	0.455	0.455	0.18
Residual	8	20.614	2.577	
Total	15	132.645	8.843	

    60 PREDICT   [PRINT=PRED,DESC,SE] K,PN[1],PN[2]

*** Predictions from regression model ***

The predictions are based on fixed values of some variates:

Variate	Fixed value	Source of value
PN[1]	0.	Mean of variate

The predictions are based on fixed values of some variates:

Variate	Fixed value	Source of value
PN[2]	-0.355E-14	Mean of variate

The predictions have been standardized by averaging
fitted values over the levels of some factors:

Factor	Weighting policy	Status of weights
BLOCK	Marginal weights	Constant over levels of other factors

    Table contains predictions followed by standard errors

Response variate: Z

K		
200	0.5322	0.0259
300	0.5123	0.0267

* Standard errors are approximate, since model is not linear

    61 STOP

******** End of Example M. Maximum of 18424 data units used at line 57 (162008 left)

SUGGESTED FURTHER WORK

1. Examine the adequacy of fit of the above logistic model.
2. Note in particular the magnitude of the residual deviance in the logistic analysis. By adopting a quasi-likelihood approach, adjust the standard errors of the estimated coefficients.
3. Compare critically the answers from the two methods of analysis presented here.

Example N

Subjective preference data on soap pads

DESCRIPTION OF DATA

Table N.1 gives data obtained during the development of a soap pad. The factors, amount of detergent, d, coarseness of pad, c, and solubility of detergents, s, were each set at two levels. There were 32 judges and the experiment was done on two days. Each judge attached a score (excellent $= 1, \ldots$, poor $= 5$) to two differently formulated pads on each of two days. For the data and several different analyses, see Johnson (1967).

THE ANALYSIS

The experimental design is rather complex, and is discussed in *App. Stat.*

Marginal means and frequencies of scores, for preliminary inspection, are readily obtained using TABULATE and CALCULATE commands.

For the main analysis we follow the approach in *App. Stat.* and ignore any repetition of judges, i.e. we compute the analysis assuming there to be 32 different pairs of judges. Inspection of residuals can be used as a check on the justification of this. Scores on the five-point scale are treated as an ordinary quantitative variable; note that there are not many extreme values (1 or 5).

The main analysis of variance table, Table N.4, *App. Stat.*, is obtained in two stages:

1. for the complete 2^3 factorial having treatments d, c and s;
2. for the days, replicates, treatments and judges (nested within treatments and replicates).

Table N.1 Subjective scores allocated to soap pads prepared in accordance with 2^3 factorial scheme. Five-point scale: 1 = excellent, 5 = poor.

Judge	Treatment	Day 1	Day 2	Judge	Treatment	Day 1	Day 2
Replicate I				Replicate II			
1	1	2	4	5	1	4	2
17	1	2	3	21	1	3	3
1	dcs	4	4	5	cs	3	4
17	dcs	4	4	21	cs	1	2
2	d	5	4	6	d	1	2
18	d	4	4	22	d	5	4
2	cs	2	1	6	dcs	3	3
18	cs	1	2	22	dcs	4	4
3	c	1	3	7	c	3	3
19	c	5	5	23	c	3	5
3	ds	3	2	7	s	4	4
19	ds	4	3	23	s	5	3
4	s	1	3	8	dc	4	4
20	s	2	3	24	dc	2	3
4	dc	3	4	8	ds	3	2
20	dc	3	3	24	ds	2	3
ReplicateIII				Replicate IV			
9	1	3	2	13	1	3	4
25	1	2	3	29	1	3	4
9	ds	4	3	13	dc	2	3
25	ds	3	3	29	dc	3	4
10	d	1	1	14	d	4	4
26	d	3	3	30	d	4	3
10	s	2	1	14	c	2	2
26	s	1	1	30	c	4	5
11	c	3	3	15	s	5	5
27	c	3	3	31	s	3	3
11	dcs	3	3	15	dcs	4	4
27	dcs	2	2	31	dcs	1	2
12	dc	3	3	16	ds	4	3
28	dc	4	4	32	ds	4	3
12	cs	1	2	16	cs	3	4
28	cs	3	3	32	cs	1	4

PROGRAM

```
 2  JOB        'Example N'
 3
 4  OPEN       'XN'; 2
 5
 6  "Data:
-7  1   i  2  4
-8  ...
-9  32  cs 1  4 :
-10
-11 We have used i rather than 1 to represent the lowest treatment level. It is
-12 possible to read 1,s,c ... as text values, but either the 1 has to be punched
-13 as '1' or a fixed format used in the READ."
14
15  UNITS      [64]
16  TEXT       NT; !T(i,s,c,cs,d,ds,dc,dcs)
17  FACTOR     [LEVELS=4] Replica; !(16(1...4))
18  &          [LEVELS=32] JUDGE
19  FACTOR     [LABELS=NT] TMT
20  READ       [CHANNEL=2] JUDGE,TMT,SCORE1,SCORE2; FREP=*,LABELS,*,*
```

Identifier	Minimum	Mean	Maximum	Values	Missing
SCORE1	1.000	2.922	5.000	64	0
SCORE2	1.000	3.125	5.000	64	0

```
21
22  "Table N.2 App. Stat. is produced by TABULATE. The INTERLEAVE option of PRINT
-23 displays means for the replicate-day combinations in the form required."
24
25  TABULATE   [CLASS=Replica]   SCORE1,SCORE2; MEANS=Day1,Day2
26  PRINT      [INTERLEAVE-1] Day1,Day2; FIELD=10; DECIMALS=3  "Table N.2 App. Stat."
```

Replica	1	2	3	4
Day1	2.875	3.125	2.563	3.125
Day2	3.250	3.187	2.500	3.563

```
27  "To form Table N.3 App. Stat., the frequency distributions for the two days are
-28 first obtained separately. and then combined by EQUATE and CALCULATE to form a
-29 single table (FD) of scores by treatment. The means for each treatment are
-30 obtained by  defining a table W of weights (1 to 5) classified by score and a
-31 table M classified by TMT; multiplying W by FD and assigning the result to M
-32 gives the marginal totals for treatments, summed over scores. The division by
-33 16 converts the totals to means."
34
35  FACTOR     [LEVELS=5] F1,F2,FS
36  CALCULATE  F1,F2 = SCORE1,SCORE2
37  TABULATE   [CLASS=F1,TMT;COUNT=CT1]
38  &          [CLASS=F2,TMT;COUNT=CT2]
39  TABLE      [CLASS=FS,TMT] FD,S1,S2
40  EQUATE     CT1,CT2; S1,S2
41  CALCULATE  FD = S1+S2
42  TABLE      [CLASS=TMT]   M
43  &          [CLASS=FS] W; !(1...5)
44  CALCULATE  M = W*FD/16
45  VARIATE    [NVALUES=8] Mean; !(#M)
46  PRINT      FD; 7; 0  & [ORIENT=ACROSS]  Mean; 7; 3       "Table N.3 App. Stat."
```

	FD							
TMT	i	s	c	cs	d	ds	dc	dcs
FS								
1	0	4	1	5	3	0	0	1
2	5	2	2	4	1	3	2	3
3	7	5	8	4	3	9	8	4
4	4	2	1	3	7	4	6	8
5	0	3	4	0	2	0	0	0

Mean	2.938	2.875	3.313	2.313	3.250	3.063	3.250	3.188

```
 47   "To get the analysis of variance for the factorial treatments, TMT has to be
-48   split into its three components, using logical expressions in CALCULATE.
-49   Genstat treats logical values numerically, representing 'false' and 'true' by
-50   0.0 and 1.0. The effect of TMT.EQ.2 is to create a vector of the same length
-51   with values of 1.0 or 0.0 corresponding to the values which are or are not
-52   equal to 2. (The final +1 ensures that the final values assigned to the factors
-53   are 1 and 2 rather than 0 or 1.)"
 54
 55   FACTOR     [LEVELS=2] S,C,D
 56   CALCULATE  S = (TMT.EQ.2 .OR. TMT.EQ.4 .OR. TMT.EQ.6 .OR. TMT.EQ.8) +1
 57   &          C = (TMT.EQ.3 .OR. TMT.EQ.4 .OR. TMT.EQ.7 .OR. TMT.EQ.8) +1
 58   &          D = (TMT.EQ.5 .OR. TMT.EQ.6 .OR. TMT.EQ.7 .OR. TMT.EQ.8) +1
 59
 60   "To ignore the effect of judges, variates and factors of length 128 combining
-61   the data for the two days should be set up:"
 62
 63   UNITS      [128]
 64   VARIATE    SCORES; !(#SCORE1,#SCORE2)
 65   FACTOR     [LEVELS=4]  REPS; !(16(1...4)2)
 66   FACTOR     [LEVELS=2]  DAYS; !(64(1,2))
 67   FACTOR     [LEVELS=32] JUDGES; !((#JUDGE)2)
 68   FACTOR     [LEVELS=2]  SS,CS,DS; !((#S)2),!((#C)2),!((#D)2)
 69   FACTOR     [LEVELS=8]  TMTS; !((#TMT)2)
 70   BLOCK      REPS*DAYS
 71   TREATMENT  SS*CS*DS
 72   ANOVA      [PRINT=AOV] SCORES                          "Table N.4 App. Stat."
```

***** Analysis of variance *****

Variate: SCORES

Source of variation	d.f.	s.s.	m.s.	v.r.
REPS stratum	3	11.648	3.883	8.33
DAYS stratum	1	1.320	1.320	2.83
REPS.DAYS stratum	3	1.398	0.466	0.43
REPS.DAYS.*Units* stratum				
SS	1	3.445	3.445	3.18
CS	1	0.008	0.008	0.01
DS	1	3.445	3.445	3.18
SS.CS	1	1.320	1.320	1.22
SS.DS	1	1.320	1.320	1.22
CS.DS	1	0.195	0.195	0.18
SS.CS.DS	1	2.258	2.258	2.08
Residual	113	122.570	1.085	
Total	127	148.930		

```
 73   "To produce the rest of the analysis shown in Table N.4, we revert to TMT."
 74
 75   BLOCK
 76   TREATMENT DAYS*((REPS*TMTS)/JUDGES)
 77   ANOVA     [PRINT=AOV] SCORES                          "Table N.4 App. Stat."
```

***** Analysis of variance *****

Variate: SCORES

Source of variation	d.f.	s.s.	m.s.	v.r.
DAYS	1	1.3203	1.3203	3.60
REPS	3	11.6484	3.8828	10.57
TMTS	7	11.9922	1.7132	4.67
DAYS.REPS	3	1.3984	0.4661	1.27
DAYS.TMTS	7	6.7422	0.9632	2.62
REPS.TMTS	21	38.0391	1.8114	4.93
DAYS.REPS.TMTS	21	10.2891	0.4900	1.33
REPS.TMTS.JUDGES	32	55.7500	1.7422	4.74
Residual	32	11.7500	0.3672	
Total	127	148.9297		

```
78
79    "'REPS.TMT.JUDGES' corresponds to 'Between Judges, within treatments, within
-80   replicates' in App. Stat.; 'Residual' to 'Days x Between Judges.
-81
-82   To produce the within-judge analysis of App. Stat.:"
83
84    CALCULATE  SCORE = SCORE1+SCORE2
85    UNITS      [32]
86    VARIATE    RESPONSE,VTM
87    FACTOR     [LEVELS=16] SESSION; !(2(1...16))
88    CALCULATE  RESPONSE$[!(1,3...31)] = SCORE$[!(3,7...63)]-SCORE$[!(1,5...61)]
89    &          RESPONSE$[!(2,4...32)] = SCORE$[!(4,8...64)]-SCORE$[!(2,6...62)]
90    &          VTM$[!(1...32)] = TMT$[!(1,3...63)]
91
92    "The design matrix is formed in a set of variates THETA[1...8]."
93
94    POINTER    [NVALUES=NT] THETA
95    VARIATE    THETA[]
96    CALCULATE  THETA[] = 0
97    FOR        I=2(1,3...31); J=2(2,4...32); K=1...32
98    CALCULATE  THETA[#VTM$[I],#VTM$[J]]$[K] = -1,1
99    ENDFOR
100
101   MODEL      RESPONSE
102   FIT        [PRINT=SUMMARY;CONSTANT=OMIT] THETA['s','c','cs','d','ds','dc','dcs']
```

***** Regression Analysis *****

*** Summary of analysis ***

	d.f.	s.s.	m.s.	v.r.
Regression	7	78.8	11.262	2.20
Residual	25	128.2	5.127	
Total	32	207.0	6.469	

Percentage variance accounted for 16.6

```
103   "This gives three of the entries of Table N.5, App. Stat."
104
105   RKEEP      RESPONSE; ESTIMATES=Estimates
106   VARIATE    Thetas; !(0,#Estimates)
107   FIT        [PRINT=SUMMARY] SESSION
```

***** Regression Analysis *****

*** Summary of analysis ***

	d.f.	s.s.	m.s.	v.r.
Regression	15	118.97	7.931	1.77
Residual	16	71.50	4.469	
Total	31	190.47	6.144	

Percentage variance accounted for 27.3

* MESSAGE: The following units have large standardized residuals:
```
              27     -2.01
              28      2.01
```

```
108   "The residual s.s. (71.50) is the 'Between pairs within sessions' s.s. of
-109  Table N.5; the s.s. Between Sessions may be obtained by subtraction."
110
111   RKEEP      ESTIMATES=EST
112   SCALAR     Emean
113   CALCULATE  Emean = EST$[1]
114   &          Adj_Mean = Emean+Thetas/2
115   &          Adj_Mean = (MEAN(Mean)/MEAN(Adj_Mean))*Adj_Mean
116   PRINT      [ORIENT=ACROSS] NT,Thetas,Mean,Adj_Mean;                      \
117              FIELD=8; DEC=2                          "Table N.6 App. Stat."
```

NT	i	s	c	cs	d	ds	dc	dcs
Thetas	0.00	1.79	1.58	-1.38	2.63	0.42	1.46	1.50
Mean	2.94	2.88	3.31	2.31	3.25	3.06	3.25	3.19
Adj_Mean	2.52	3.42	3.32	1.83	3.84	2.73	3.25	3.28

118 STOP

******** End of Example N. Maximum of 22846 data units used at line 99 (157586 left)

SUGGESTED FURTHER WORK

1. Investigate the large residuals identified in the above analysis.
2. Define a response variable Z,

$$Z = \begin{cases} 0 & \text{if score less than } k \\ 1 & \text{otherwise} \end{cases}$$

   where $1 < k \leq 5$, and fit a logistic model.
3. Compare your conclusions from (2) with those from the analysis of the straight score.
4. Consider how the logistic models of (2) for $k = 2, \ldots, 5$ may be combined.

Example O

Atomic weight of iodine

DESCRIPTION OF DATA

Table O.1 gives ratios of reacting weight of iodine and silver obtained, in an accurate determination of atomic weight of iodine, using five batches of silver A, B, C, D, E and two of iodine, I, II (Baxter and Landstredt, 1940; Brownlee, 1965). Silver batch C is a repurification of batch B, which in turn is a repurification of batch A. In these data 1.176 399 has been subtracted from all values.

Table O.1 Ratios of reacting weight with 1.176 399 subtracted $\times 10^6$

| | Iodine batch | |
Silver batch	I	II
A	23, 26	0, 41, 19
B	42, 42	24, 14
C	30, 21, 38	
D	50, 51	62
E	56	

THE ANALYSIS

This example illustrates in very simple form aspects arising widely in the analysis of unbalanced data. Because of the lack of balance it is necessary to proceed by fitting a sequence of models and differencing the residual sums of squares from two models to obtain the appropriate sum of squares (s.s.) for testing columns (adjusted for rows) or rows (adjusted for columns). If Y_{ijk} denotes the kth observation in row i and column j, the sequence of models suggested is:

Model I $\quad$ E$(Y_{ijk}) = \mu$; homogeneity

Model II$_1$ E$(Y_{ijk}) = \mu + \alpha_i$; pure row effects

Model II_2 $E(Y_{ijk}) = \mu + \beta_j$; pure column effects
Model II_{12} $E(Y_{ijk}) = \mu + \alpha_i + \beta_j$; additivity (no interaction)
Model III $E(Y_{ijk}) = \mu_{ij} = \mu + \alpha_i + \beta_j + \gamma_{ij}$; arbitrary means.

(O.1, *Appl. Stat.*)

We fit model II_1 (silver) and then ADD IODINE for model II_{12}, printing the accumulated s.s. to obtain the s.s. for iodine (adjusted for silver). Likewise model II_2 (iodine) is fitted followed by ADD SILVER and ADD SILVER.IODINE. This gives all the s.s. required for the analysis of variance of the unbalanced two-way layout, Table O.2(b), *App. Stat.* The unadjusted and adjusted silver means, Table O.3(a), *App. Stat.*, are given by PREDICT SILVER having fitted model II_1 and II_{12}, respectively.

PROGRAM

```
 2  JOB        'Example O'
 3
 4  OPEN       'XO'; 2
 5
 6  "The compact presentation of the data in Table O.1 App. Stat. is not convenient
-7  and the simplest solution is to give full details (Silver Batch, Iodine Batch
-8  and Ratio) for each unit. No information is required for non-existent units:
-9  A 1 23
-10 ...
-11 E 1 56 : "
 12
 13 UNITS      [16]
 14 TEXT       NS; !T(A,B,C,D,E)
 15 FACTOR     [LABELS=NS]  SILVER
 16 FACTOR     [LEVELS=2]   IODINE
 17 READ       [CHANNEL=2] SILVER,IODINE,RATIO; FREP=LABELS,LEVELS,*
```

	Identifier	Minimum	Mean	Maximum	Values	Missing
	RATIO	0.00	33.69	62.00	16	0

```
 18
 19 "For unbalanced data, there is no unique sums of squares for a given effect,
-20 and the order of fitting becomes relevant. If we fit SILVER+IODINE, the a.o.v.
-21 table shows the sums of squares for
-22    SILVER, ignoring IODINE     and
-23    IODINE, adjusted for SILVER
-24 Having done this, we can start again and fit IODINE+SILVER, to get the s.s. for
-25    IODINE, ignoring SILVER     and
-26    SILVER, adjusted for IODINE
-27 The interaction term IODINE.SILVER can then be added."
 28
 29 MODEL      RATIO
 30 TERMS      SILVER*IODINE
 31 FIT        "Model II1"  [PRINT=SUMMARY] SILVER
```

***** Regression Analysis *****

*** Summary of analysis ***

	d.f.	s.s.	m.s.	v.r.
Regression	4	2572.	643.1	4.20
Residual	11	1683.	153.0	
Total	15	4255.	283.7	
Change	-4	-2572.	643.1	4.20

Percentage variance accounted for 46.1

* MESSAGE: The following units have high leverage:
           16   1.00

  32 PREDICT [PRINT=DESCRIPTIONS,PREDICTIONS,SE] SILVER   "Table 0.3(a) App. Stat."

*** Predictions from regression model ***

      Table contains predictions followed by standard errors

  Response variate: RATIO

        SILVER
          A   21.80   5.53
          B   30.50   6.18
          C   29.67   7.14
          D   54.33   7.14
          E   56.00   12.37

  33 "These are the estimated silver batch means, unadjusted for iodine,
 -34 when model II1 is fitted."
  35
  36 ADD   [PRINT=ACCUMULATED] IODINE

***** Regression Analysis *****

*** Accumulated analysis of variance ***

Change	d.f.	s.s.	m.s.	v.r.
+ SILVER	4	2572.3	643.1	4.19
+ IODINE	1	150.0	150.0	0.98
Residual	10	1533.2	153.3	
Total	15	4255.4	283.7	

  37 FIT   "Model II2" [PRINT=SUMMARY] IODINE

***** Regression Analysis *****

*** Summary of analysis ***

	d.f.	s.s.	m.s.	v.r.
Regression	1	473.	473.2	1.75
Residual	14	3782.	270.2	
Total	15	4255.	283.7	
Change	-1	-473.	473.2	1.75

Percentage variance accounted for 4.8

* MESSAGE: The following units have large standardized residuals:
           15   2.35

  38 ADD   "Model II12" [PRINT=SUMMARY] SILVER ·

***** Regression Analysis *****

*** Summary of analysis ***

	d.f.	s.s.	m.s.	v.r.
Regression	5	2722.	544.5	3.55
Residual	10	1533.	153.3	
Total	15	4255.	283.7	
Change	-4	-2249.	562.3	3.67

Percentage variance accounted for 46.0

* MESSAGE: The following units have large standardized residuals:
          4   2.07

* MESSAGE: The following units have high leverage:
         16   1.00

```
 39   PREDICT   [PRINT=DESCRIPTIONS,PREDICTIONS,SE; ADJUST=EQUAL]                    \
 40             SILVER                                    "Table O.3(a) App. Stat."
```

*** Predictions from regression model ***

The predictions have been standardized by averaging
fitted values over the levels of some factors:

```
          Factor   Weighting policy   Status of weights
          IODINE       Equal weights   Constant over levels of other factors
```

    Table contains predictions followed by standard errors

 Response variate: RATIO

```
          SILVER
               A      22.52      5.59
               B      30.50      6.19
               C      26.05      8.03
               D      53.13      7.25
               E      52.38     12.91
```

```
 41   "These are the estimated silver batch means, adjusted for iodine,
-42    when model II12 is fitted."
 43
 44   ADD       "Model III"   [PRINT=ACCUMULATED] SILVER.IODINE
```

* MESSAGE: Term SILVER.IODINE cannot be fully included in the model
  because 2 parameters are aliased with terms already in the model

***** Regression Analysis *****

*** Accumulated analysis of variance ***

Change	d.f.	s.s.	m.s.	v.r.
+ IODINE	1	473.2	473.2	3.63
+ SILVER	4	2249.1	562.3	4.32
+ SILVER.IODINE	2	491.5	245.8	1.89
Residual	8	1041.7	130.2	
Total	15	4255.4	283.7	

```
 45   "Tables O2(a) and (b) can be constructed from the output.
-46    There is no evidence of significant Iodine or Iodine.Silver effect, but the
-47    within cells s.s. on 8 d.f. (1041.7) should be used to calculate the s.e.s of
-48    the estimates."
 49
 50   STOP
```

******** End of Example O. Maximum of 8796 data units used at line 40 (40564 left)

SUGGESTED FURTHER WORK

Note those observations indicated as having large residuals or high
leverage and carry out any further analysis that might be appropriate.

Example P

Multifactor experiment on a nutritive medium

DESCRIPTION OF DATA

Fedorov, Maximov and Bogorov (1968) obtained the data in Table P.1 from an experiment on the composition of a nutritive medium for green sulphur bacteria *Chlorobrium thiosulphatophilum*. The bacteria were grown under constant illumination at a temperature of 25–30 °C: the yield was determined during the stationary phase of growth. Each factor was at two levels, with each level used 8 times. Subject to this, the factor levels were randomised.

THE ANALYSIS

Although this problem raises some difficult statistical issues, computationally it is straightforward, involving a series of multiple regressions, readily fitted by Genstat.

The analysis in *App. Stat.* starts by fitting a main-effects model containing all 10 component variables $x_1, \ldots, x_{10}$. The residual mean square from this model is appreciably greater than the external estimate of variance (3.8^2) and so models containing cross-product terms are fitted but, with only 16 observations, the choice of possible terms is restricted. A model is built up from a small number of main effects, choosing those found to be appreciable in the initial analysis, plus their interactions. This leads to a reasonably well fitting model with four main effects, x_6, x_7, x_8, x_9, and two cross-products, $x_6 x_8$ and $x_7 x_8$. There may be other models fitting equally well.

It should be noted that some orthogonality exists between the variables.

Table P.1 Yields of bacteria

					Factors						
	x_1 NH_4Cl	x_2 KH_2PO_4	x_3 $MgCl_2$	x_4 $NaCl$	x_5 $CaCl_2$	x_6 $Na_2S\cdot$ $9H_2O$	x_7 $Na_2S_2O_3$	x_8 $NaHCO_3$	x_9 $FeCl_3$	x_{10} micro elements	Y Yield
Levels $\{$ + / −	1500 / 500	450 / 50	900 / 100	1500 / 500	350 / 50	1500 / 500	5000 / 1000	5000 / 1000	125 / 25	15 / 5	
1	−	+	+	+	−	+	−	+	−	+	14.0
2	−	−	+	+	−	+	+	−	−	+	4.0
3	+	−	−	+	+	+	−	−	−	−	7.0
4	−	−	+	−	+	+	−	+	+	+	24.5
5	+	−	+	+	+	+	+	−	−	−	14.5
6	+	−	+	−	+	+	+	+	+	+	71.0
7	−	−	−	−	−	−	−	−	−	−	15.5
8	+	+	−	+	+	−	−	+	+	−	18.0
9	−	+	−	+	−	−	+	−	−	+	17.0
10	+	+	+	+	+	−	−	−	+	−	13.5
11	−	+	+	−	+	−	+	+	+	+	52.0
12	+	+	+	−	−	−	+	+	−	−	48.0
13	+	+	−	−	+	−	+	−	+	−	24.0
14	−	+	−	−	−	+	−	−	+	−	12.0
15	+	−	−	−	−	−	−	+	+	+	13.5
16	−	−	−	+	−	+	+	+	−	+	63.0

All the concentrations are given in mg/l, with the exception of factor 10, whose central level (10 ml of solution of micro-element per litre of medium) corresponds to 10 times the amount of micro-element in Larsen's medium. The yield has a standard error of 3.8.

PROGRAM

```
   2   JOB       'Example P'
   3
   4   OPEN      'XP'; 2
   5
   6   "The actual levels need not be declared; the factor values are represented
  -7   by + and - :
  -8   -+++-+-+-+   14.0
  -9   ...
 -10   ---+-+++-+   63.0 :
 -11   and read using fixed format."
  12
  13   UNITS     [16]
  14   TEXT      NF; !T('-','+')
  15   FACTOR    [LABELS=NF]                                                    \
  16             NH4Cl,KH2PO4,MgCl2,NACl,CaCl2,Na2S9H2O,Na2S2O3,NaHCO3,FeCl3,Micros
  17   READ      [CHANNEL=2;LAYOUT=FIXED;SKIP=*]  NH4Cl,KH2PO4,MgCl2,NACl,CaCl2,  \
  18             Na2S9H2O,Na2S2O3,NaHCO3,FeCl3,Micros,YIELD; FIELD=10(1),7; FREP=LABELS
```

Identifier	Minimum	Mean	Maximum	Values	Missing
YIELD	4.00	25.72	71.00	16	0

```
  19
  20   CALCULATE X[1...10] =                                                    \
  21        (NH4Cl,KH2PO4,MgCl2,NACl,CaCl2,Na2S9H2O,Na2S2O3,NaHCO3,FeCl3,Micros)
  22
  23   "It is never sensible to seek a 'best' model by automatic selection of subsets
 -24   of the explanatory variables (even when there are sufficient data) and Genstat
 -25   makes no provision for this. Instead, the STEP command may be used to
 -26   investigate data sets for which little prior knowledge is available. It can be
 -27   used for forward selection or backward elimination.
 -28
 -29   We follow the approach described in App. Stat. and first fit all 10 variables."
  30
  31   MODEL     YIELD
  32   TERMS     X[]
  33   FIT       [PRINT=ACCUMULATED,SUMMARY,ESTIMATES,MODEL] X[]
```

***** Regression Analysis *****

Response variate: YIELD
     Fitted terms: Constant, X[1], X[2], X[3], X[4], X[5], X[6], X[7], X[8], X[9], X[10]

*** Summary of analysis ***

	d.f.	s.s.	m.s.	v.r.
Regression	10	4916.	491.6	1.61
Residual	5	1528.	305.7	
Total	15	6444.	429.6	
Change	-10	-4916.	491.6	1.61

Percentage variance accounted for 28.8

*** Estimates of regression coefficients ***

	estimate	s.e.	t
Constant	-28.8	45.4	-0.63
X[1]	-6.4	14.0	-0.46
X[2]	-3.2	13.6	-0.24
X[3]	-4.4	10.8	-0.41
X[4]	-6.0	13.8	-0.43
X[5]	5.0	14.4	0.34
X[6]	3.7	11.9	0.31
X[7]	26.3	11.5	2.30
X[8]	28.5	12.3	2.32
X[9]	3.0	17.8	0.17
X[10]	-10.1	17.7	-0.57

*** Accumulated analysis of variance ***

Change	d.f.	s.s.	m.s.	v.r.
+ X[1]	1	3.5	3.5	0.01
+ X[2]	1	13.1	13.1	0.04
+ X[3]	1	319.5	319.5	1.05
+ X[4]	1	749.4	749.4	2.45
+ X[5]	1	22.9	22.9	0.07
+ X[6]	1	14.0	14.0	0.05
+ X[7]	1	1732.3	1732.3	5.67
+ X[8]	1	1951.4	1951.4	6.38
+ X[9]	1	9.1	9.1	0.03
+ X[10]	1	100.2	100.2	0.33
Residual	5	1528.5	305.7	
Total	15	6444.0	429.6	

```
  34  "X[7] and X[8] are clearly of greatest importance."
  35
  36  CALCULATE X78 = X[7]*X[8]
  37  TERMS     X[ ],X78
  38  FIT       X[7,8],X78
```

***** Regression Analysis *****

Response variate: YIELD
      Fitted terms: Constant, X[7], X[8], X78

*** Summary of analysis ***

	d.f.	s.s.	m.s.	v.r.
Regression	3	5791.8	1930.60	35.52
Residual	12	652.2	54.35	
Total	15	6444.0	429.60	
Change	-3	-5791.8	1930.60	35.52

Percentage variance accounted for 87.3

*** Estimates of regression coefficients ***

	estimate	s.e.	t
Constant	41.7	18.4	2.27
X[7]	-35.2	11.7	-3.02
X[8]	-32.6	11.7	-2.80
X78	38.12	7.37	5.17

```
  39  "To see if any of the other variables are useful, we could now use STEP:
 -40  FOR       [NTIMES=8]
 -41    STEP    [OUTRATIO=*] X[1...6,9,10]
 -42  ENDFOR
 -43  If this is done, it will be found that only X[9] is of interest. The
 -44  output is not reproduced here."
  45
  46  CALCULATE X68 = X[6]*X[8]
  47  TERMS     X[6...9],X68,X78
  48  FIT       X[6...9],X68,X78
```

***** Regression Analysis *****

Response variate: YIELD
      Fitted terms: Constant, X[6], X[7], X[8], X[9], X68, X78

*** Summary of analysis ***

	d.f.	s.s.	m.s.	v.r.
Regression	6	6272.1	1045.35	54.73

```
Residual        9        171.9        19.10
Total          15       6444.0       429.60

Change         -6      -6272.1      1045.35      54.73
```

Percentage variance accounted for 95.6

*** Estimates of regression coefficients ***

```
                   estimate        s.e.         t
Constant             69.3         15.7        4.40
X[6]                -24.87         6.94       -3.59
X[7]                -33.62         6.94       -4.85
X[8]                -61.82         9.54       -6.48
X[9]                  6.52         2.42        2.69
X68                  18.37         4.37        4.20
X78                  38.12         4.37        8.72
```

```
 49  "Note that the standard errors printed here are computed using the residual mean
-50  square (19.10); those in Table P.3 App. Stat. are computed using the external
-51  standard error 3.8 which is slightly smaller.
-52
-53  Covariances of the estimates, and residuals, are obtained by:"
 54
 55  RKEEP      YIELD; FITTED=FITTED; VCOVARIANCE=VCM
 56  PRINT      VCM; FIELD=10; DECIMALS=4
```

```
                VCM

Constant  247.4808
   X[6]    -74.3820   48.1187
   X[7]    -74.3820    0.3673   48.1187
   X[8]  -133.3366   42.6089   42.6089   91.0950
   X[9]   -11.0196    1.4693    1.4693   -1.4693    5.8771
    X68    42.9763  -28.6508    0.0000  -28.6508    0.0000   19.1006
    X78    42.9763    0.0000  -28.6508  -28.6508    0.0000    0.0000   19.1006

           Constant      X[6]      X[7]      X[8]      X[9]       X68       X78
```

```
 57  CALCULATE RESIDUAL = YIELD-FITTED
 58  PRINT      YIELD,FITTED,RESIDUAL; FIELD=10; DECIMALS=1
```

```
    YIELD    FITTED   RESIDUAL
    14.0     18.6      -4.6
     4.0     10.0      -6.0
     7.0      5.5       1.5
    24.5     25.1      -0.6
    14.5     10.0       4.5
    71.0     67.7       3.3
    15.5     12.0       3.5
    18.0     13.2       4.8
    17.0     16.5       0.5
    13.5     18.5      -5.0
    52.0     55.8      -3.8
    48.0     49.3      -1.3
    24.0     23.0       1.0
    12.0     12.0       0.0
    13.5     13.2       0.3
    63.0     61.2       1.8
```

```
 59  "and a plot of residuals against normal order statistics may be produced by
-60  RCHECK as in Example G."
 61
 62  STOP
```

******** End of Example P. Maximum of 11098 data units used at line 57 (38262 left)

SUGGESTED FURTHER WORK

1. Consider what other models might be fitted. In particular, examine whether the regression of log yield on log concentrations gives a better model.
2. Consider and implement graphical methods for choosing the interactions for inclusion.

Example Q
Strength of cotton yarn

DESCRIPTION OF DATA

An experiment was done with the objects of estimating:

1. the difference in mean strength of two worsted yarns produced by slightly different processes, and
2. the variation of strength between and within bobbins for yarns of this type.

For each yarn a considerable number of bobbins were produced and six bobbins selected at random. From each of these, four short lengths were chosen at random for strength testing. The breaking loads are given in Table Q.1.

Table Q.1 Breaking loads (oz)

Bobbin	1	2	3	4	5	6
Yarn A	15.0	15.7	14.8	14.9	13.0	15.9
	17.0	15.6	15.8	14.2	16.2	15.6
	13.8	17.6	18.2	15.0	16.4	15.0
	15.5	17.1	16.0	12.8	14.8	15.5
Yarn B	18.2	17.2	15.2	15.6	19.2	16.2
	16.8	18.5	15.9	16.0	18.0	15.9
	18.1	15.0	14.5	15.2	17.0	14.9
	17.0	16.2	14.2	14.9	16.9	15.5

THE ANALYSIS

Preliminary analysis involves inspection of means and standard deviations for each of the six bobbins within yarn A, and within yarn B. These are obtained using TABULATE.

The nested analysis of variance, Table Q.4 *App. Stat.*, is constructed from separate analysis of variance tables which are computed as follows:

1. between yarns;
2. within yarn A;
3. within yarn B.

Components of variance between bobbins and within bobbins are estimated by REML. Note however that due to the balanced nature of the data, identical estimates are obtained by equating observed and expected mean squares in the analysis of variance table (*App. Stat.* p. 134).

PROGRAM

```
  2   JOB       'Example Q'
  3
  4   OPEN      'XQ'; 2
  5
  6   "Data:
 -7   15.0 15.7 14.8 14.9 13.0 15.9
 -8   ...
 -9   17.0 16.2 14.2 14.9 16.9 15.5 :
-10
-11   Note:
-12   1. The use of a constant 4 in the GENERATE command, to indicate the four
-13      observations on each combination of Bobbin and Yarn
-14   2. The use of TABULATE to get the means and variances for each cell. The table
-15      for means is declared with margins, as required for Table Q.2, App. Stat."
 16
 17   UNITS     [48]
 18   TEXT      NY; !T(A,B)
 19   FACTOR    [LEVELS=6]  BOBBIN
 20   FACTOR    [LABELS=NY] YARN
 21   GENERATE  YARN,4,BOBBIN
 22   READ      [CHANNEL=2] LOAD
```

Identifier	Minimum	Mean	Maximum	Values	Missing
LOAD	12.80	15.91	19.20	48	0

```
 23
 24   TABULATE  [CLASS=YARN,BOBBIN] LOAD; VARIANCES=MSQ
 25   &         [MARGINS=YES]       LOAD; MEANS=MBL
 26   PRINT     MBL; FIELD=8; DECIMALS=2                    "Table Q.2 App. Stat."
```

MBL

BOBBIN	1	2	3	4	5	6	Margin
YARN							
A	15.32	16.50	16.20	14.23	15.10	15.50	15.48
B	17.53	16.73	14.95	15.43	17.78	15.62	16.34
Margin	16.43	16.61	15.58	14.83	16.44	15.56	15.91

```
 27   &         MSQ; FIELD=8; DECIMALS=2                    "Table Q.3 App. Stat."
```

MSQ

BOBBIN	1	2	3	4	5	6
YARN						
A	1.76	1.01	2.05	1.03	2.47	0.14
B	0.53	2.21	0.58	0.23	1.15	0.32

```
28  BLOCK      YARN
29  TREATMENT  YARN/BOBBIN
30  ANOVA      [PRIN=AOV]  LOAD                        "Table Q.4 App. Stat."
```

***** Analysis of variance *****

Variate: LOAD

Source of variation	d.f.	s.s.	m.s.	v.r.
YARN	1	8.927	8.927	7.96
YARN.BOBBIN	10	40.779	4.078	3.64
Residual	36	40.383	1.122	
Total	47	90.088		

```
31  "Estimates of the components of variance are obtained by:"
32
33  VCOMPONENTS  YARN/BOBBIN
34  REML       [PRINT=COMPONENTS]  LOAD
```

*** Estimated Components of Variance ***

		s.e.
YARN	0.2020	0.5315
YARN.BOBBIN	0.7390	0.4607
units	1.122	0.2644

```
35  "The last two lines above give the components of variance for
-36  'between bobbins within yarns' and 'within bobbins'.
-37
-38  For the within-yarn terms, we restrict the analysis to each yarn separately:"
39
40  RESTRICT  LOAD; YARN.EQ.1
41  ANOVA     [PRIN=AOV]  LOAD                         "Table Q.4 App. Stat."
```

***** Analysis of variance *****

Variate: LOAD

Source of variation	d.f.	s.s.	m.s.	v.r.
YARN.BOBBIN	5	13.210	2.642	1.88
Residual	18	25.355	1.409	
Total	23	38.565		

```
42  RESTRICT  LOAD; YARN.EQ.2
43  ANOVA     [PRIN=AOV]  LOAD                         "Table Q.4 App. Stat."
```

***** Analysis of variance *****

Variate: LOAD

Source of variation	d.f.	s.s.	m.s.	v.r.
YARN.BOBBIN	5	27.5688	5.5138	6.60
Residual	18	15.0275	0.8349	
Total	23	42.5963		

```
44  STOP
```

******** End of Example Q. Maximum of 10756 data units used at line 34 (169676 left)

SUGGESTED FURTHER WORK

Examine residuals and check on the assumption of normality.

Example R

Biochemical experiment on the blood of mice

DESCRIPTION OF DATA*

In an experiment on the effect of treatments A and B on the amount of substance S in mice's blood, it was not practicable to use more than 4 mice on any one day. The treatments formed a 2×2 system:

A_0: A absent, B_0: B absent
A_1: A present, B_1: B present

The mice used on one day were all of the same sex. The data are given in Table R.1 (Cox, 1958, §7.4).

Table R.1 Amount of substance S

Day	1	Male	A_0B_1 4.8	A_1B_1 6.8	A_0B_0 4.4	A_1B_0 2.8
	2	Male	A_0B_0 5.3	A_1B_0 3.3	A_0B_1 1.9	A_1B_1 8.7
	3	Female	A_1B_1 7.2	A_0B_1 4.3	A_0B_0 5.3	A_1B_0 7.0
	4	Male	A_0B_0 1.8	A_1B_1 4.8	A_1B_0 2.6	A_0B_1 3.1
	5	Female	A_1B_1 5.1	A_0B_0 3.7	A_1B_0 5.9	A_0B_1 6.2
	6	Female	A_1B_0 5.4	A_0B_1 5.7	A_1B_1 6.7	A_0B_0 6.5
	7	Male	A_0B_1 6.2	A_1B_1 9.3	A_0B_0 5.4	A_1B_0 6.9
	8	Female	A_0B_0 5.2	A_1B_1 7.9	A_1B_0 6.8	A_0B_1 7.9

THE ANALYSIS

We note that each treatment combination (A_0B_0, A_1B_0, A_0B_1, A_1B_1) is tested on each day, but male rats are used on days 1, 2, 4 and 7 and

*Fictitious data based on a real investigation.

female rats on days 3, 5, 6 and 8. Thus comparisons between treatments within males or females are independent of systematic differences between days, but not so for any comparison of treatments across the sexes. Two different error variances are involved, demanding separate estimation.

It is helpful for initial inspection to rearrange the data in systematic order grouping by sex and by treatment as in Table R.2, *App. Stat.*; this table, with the exception of the overall means, is produced using TABULATE and COMBINE commands.

The basic analysis of variance for the split-plot design is achieved easily using the ANOVA command.

PROGRAM

```
 2  JOB        'Example R'
 3
 4  OPEN       'XR'; 2
 5
 6  "Data:
-7  A0 B1 4.8  A1 B1 6.8  A0 B0 4.4  A1 B0 2.8
-8  ...
-9  A0 B0 5.2  A1 B1 7.9  A1 B0 6.8  A0 B1 7.9  : "
10
11  UNITS      [32]
12  TEXT       NA; !T(A0,A1)
13   &         NB; !T(B0,B1)
14  FACTOR     [LABELS=NA]  A
15   &         [LABELS=NB]  B
16  FACTOR     [LEVELS=8] DAY; !(4(1...8))
17  FACTOR     [LABELS=!T(Male,Female)] SEX; !(4(1,1,2,1,2,2,1,2))
18  READ       [CHANNEL=2]  A,B,S; FREP=LABELS
```

Identifier	Minimum	Mean	Maximum	Values	Missing
S	1.800	5.466	9.300	32	0

```
19
20  "To display the data in systematic order, we first form a combined
-21  treatment factor AB and tabulate the scores by DAY and AB."
22
23  FACTOR     [LABELS=!T(A0B0,A1B0,A0B1,A1B1)] AB
24  CALCULATE  AB = A+2*B-2
25  TABLE      [CLASS=DAY,AB;MARGIN=YES] MS
26  TABULATE   S; MEANS=MS
27  PRINT      MS; FIELD=8; DECIMALS=1
```

	MS				
AB	A0B0	A1B0	A0B1	A1B1	Margin
DAY					
1	4.4	2.8	4.8	6.8	4.7
2	5.3	3.3	1.9	8.7	4.8
3	5.3	7.0	4.3	7.2	5.9
4	1.8	2.6	3.1	4.8	3.1
5	3.7	5.9	6.2	5.1	5.2
6	6.5	5.4	5.7	6.7	6.1
7	5.4	6.9	6.2	9.3	6.9
8	5.2	6.8	7.9	7.9	6.9
Margin	4.7	5.1	5.0	7.1	5.5

```
28  "The information for males and females can be extracted from this table
-29  by the COMBINE directive. Factors for days (male) and days (female), and
-30  tables classified by these and AB are defined separately."
31
32  FACTOR     [LABELS=!T('1','2','4','7')] DAYM
```

```
33    &          [LABELS=!T('3','5','6','8')] DAYF
34    TABLE      [CLASS=DAYM,AB;MARGIN=Y] Male
35    &          [CLASS=DAYF,AB;MARGIN=Y] Female
36
37    "COMBINE can now be asked to copy the appropriate slices from the full table
-38   to the new tables by specifiying the 'oldpositions' in the 'olddimension'
-39   and the 'newdimension' that they are to occupy."
40
41    COMBINE    [OLDSTRUCTURE=MS; NEWSTRUCTURE=Male]                              \
42               OLDDIMENSION=DAY; NEWDIMENSION=DAYM; OLDPOSITION=!(1,2,4,7)
43    COMBINE    [OLDSTRUCTURE=MS; NEWSTRUCTURE=Female]                           \
44               OLDDIMENSION=DAY; NEWDIMENSION=DAYF; OLDPOSITION=!(3,5,6,8)
45    MARGIN     Male,Female; METHOD=MEAN
46    PRINT      Male,Female; FIELD=8; DECIMALS=1                "Table R.2 App. Stat."
```

```
         Male
    AB   A0B0    A1B0    A0B1    A1B1   Margin
  DAYM
    1    4.4     2.8     4.8     6.8     4.7
    2    5.3     3.3     1.9     8.7     4.8
    4    1.8     2.6     3.1     4.8     3.1
    7    5.4     6.9     6.2     9.3     6.9

Margin   4.2     3.9     4.0     7.4     4.9

         Female
    AB   A0B0    A1B0    A0B1    A1B1   Margin
  DAYF
    3    5.3     7.0     4.3     7.2     5.9
    5    3.7     5.9     6.2     5.1     5.2
    6    6.5     5.4     5.7     6.7     6.1
    8    5.2     6.8     7.9     7.9     6.9

Margin   5.2     6.3     6.0     6.7     6.1
```

```
47    "The analysis of variance is straightforward:"
48
49    BLOCK      DAY
50    TREATMENT  A*B*SEX
51    ANOVA      [PRINT=AOV] S
```

***** Analysis of variance *****

Variate: S

Source of variation	d.f.	s.s.	m.s.	v.r.
DAY stratum				
SEX	1	10.928	10.928	1.80
Residual	6	36.332	6.055	4.54
DAY.*Units* stratum				
A	1	11.883	11.883	8.90
B	1	10.465	10.465	7.84
A.B	1	5.528	5.528	4.14
A.SEX	1	0.813	0.813	0.61
B.SEX	1	1.950	1.950	1.46
A.B.SEX	1	8.508	8.508	6.37
Residual	18	24.026	1.335	
Total	31	110.432		

```
52    "This produces the basic details for constructing Table R.3, App. Stat.
-53   These can be extracted using AKEEP, and put together for display."
54
55    TEXT       AV; !T('Between sexes','Between days within sex',' Between Days',  \
56               'A','B','AxB','   Treatments','A x Sex','B x Sex','A x B x Sex',   \
57               '   Treatments x Sex','Residual','    Total')
58    AKEEP      DAY+SEX+A*B*SEX; SS=SS[1...8]; DF=DF[1...8]
59    VARIATE    [NVALUES=13] df,ss,ms
60    CALCULATE  (df,ss)$[1,2,4,5,6,8,9,10] = (DF,SS)[2,1,3,4,7,5,6,8]
```

```
61    &         XD,XS = SUM(df,ss)
62    &         (df,ss)$[3] = (DF,SS)[1]+(DF,SS)[2]
63    &         (df,ss)$[7] = (DF,SS)[3]+(DF,SS)[4]+(DF,SS)[7]
64    &         (df,ss)$[11]= (DF,SS)[5]+(DF,SS)[6]+(DF,SS)[8]
65    &         df$[13] = NVAL(S)-1 & ss$[13] = df$[13]*VAR(S)
66    &         (df,ss)$[12] = (df,ss)$[13]-XD,XS
67    &         ms = ss/df
68    PRINT     [INDENTATION=10] AV,df,ss,ms; FIELD=30,6,10,10; DECIMALS=0,0,2,2;    \
69              JUSTIFICATION=LEFT,3(RIGHT)                    "Table R.3 App. Stat."
```

AV	df	ss	ms
Between sexes	1	10.93	10.93
Between days within sex	6	36.33	6.06
Between Days	7	47.26	6.75
A	1	11.88	11.88
B	1	10.47	10.47
AxB	1	5.53	5.53
Treatments	3	27.88	9.29
A x Sex	1	0.81	0.81
B x Sex	1	1.95	1.95
A x B x Sex	1	8.51	8.51
Treatments x Sex	3	11.27	3.76
Residual	18	24.03	1.33
Total	31	110.43	3.56

```
70
71    STOP
```

******** End of Example R. Maximum of 10732 data units used at line 64 (38628 left)

SUGGESTED FURTHER WORK

1. Examine the assumption of normality.
2. See whether a transformation will reduce the sex $\times$ B $\times$ A interaction and hence simplify the conclusions.
3. Consider treating days as a random effect. What justification, if any, is there for treating days as random?
4. Estimate a standard error for the comparison of a particular treatment combination (A_1B_1, say) across the sexes.

Example S
Voltage regulator performance

DESCRIPTION OF DATA

Voltage regulators fitted to private motor cars were required to operate within the range of 15.8 to 16.4 volts, and the following investigation (Desmond, 1954) was conducted to estimate the pattern of variability encountered in production. The normal procedure was for a regulator from the production line to be passed to one of a number of setting stations, where the regulator was adjusted on a test rig. These regulators then passed to one of four testing stations, where the regulator was tested, and if found to be unsatisfactory, it was passed down the production line to be reset. For the data of Table S.1, a random sample of four setting stations took part, and a number of regulators from each setting station were passed through each testing station. One special aspect of interest concerned the percentage of regulators that would be unsatisfactory were the mean kept constant at 16.1.

THE ANALYSIS

Inspection across the readings for each regulator shows close consistency, except for regulator J_2 and to a lesser extent B_2. We compute a separate two-way (testing stations × regulators) analysis of variance for each of the ten setting stations A, ..., K. Due to the anomalous behaviour of regulator J_2 subsequent analysis is done both with and without J_2.

An overall analysis of variance, Table S.3, *App. Stat.*, is obtained by the ANOVA command having defined the treatments structure SS∗TS + SS/RG, specifying regulators nested within setting stations.

Components of variance between setting stations and between regulators are estimated using REML. The estimates differ slightly from those obtained by equating observed and expected mean squares, as in *App. Stat.*, due to the slight imbalance in the data. Then assuming normality, approximate confidence limits for the percentage of regulators can be calculated as discussed in *App. Stat.*

Table S.1 Regulator voltages

Setting station	Regulator number	Testing station			
		1	2	3	4
A	1	16.5	16.5	16.6	16.6
	2	15.8	16.7	16.2	16.3
	3	16.2	16.5	15.8	16.1
	4	16.3	16.5	16.3	16.6
	5	16.2	16.1	16.3	16.5
	6	16.9	17.0	17.0	17.0
	7	16.0	16.2	16.0	16.0
	11	16.0	16.0	16.1	16.0
B	1	16.0	16.1	16.0	16.1
	2	15.4	16.4	16.8	16.7
	3	16.1	16.4	16.3	16.3
	4	15.9	16.1	16.0	16.0
C	1	16.0	16.0	15.9	16.3
	2	15.8	16.0	16.3	16.0
	3	15.7	16.2	15.3	15.8
	4	16.2	16.4	16.4	16.6
	5	16.0	16.1	16.0	15.9
	6	16.1	16.1	16.1	16.1
	10	16.1	16.0	16.1	16.0

Setting station	Regulator number	Testing station			
		1	2	3	4
F	1	16.1	16.0	16.0	16.2
	2	16.5	16.1	16.5	16.7
	3	16.2	17.0	16.4	16.7
	4	15.8	16.1	16.2	16.2
	5	16.2	16.1	16.4	16.2
	6	16.0	16.2	16.2	16.1
	11	16.0	16.0	16.1	16.0
G	1	15.5	15.5	15.3	15.6
	2	16.0	15.6	15.7	16.2
	3	16.0	16.4	16.2	16.2
	4	15.8	16.5	16.2	16.2
	5	15.9	16.1	15.9	16.0
	6	15.9	16.1	15.8	15.7
	7	16.0	16.4	16.0	16.0
	12	16.1	16.2	16.2	16.1
H	1	15.5	15.6	15.4	15.8
	2	15.8	16.2	16.0	16.2
	3	16.2	15.4	16.1	16.3
	4	16.1	16.2	16.0	16.1
	5	16.1	16.2	16.3	16.2
	10	16.1	16.1	16.0	16.1

Table S.1 (*cont.*)

Setting station	Regulator number	Testing station			
		1	2	3	4
D	1	16.1	16.0	16.0	16.1
	2	16.0	15.9	16.2	16.0
	3	15.7	15.8	15.7	15.7
	4	15.6	16.4	16.1	16.2
	5	16.0	16.2	16.1	16.1
	6	15.7	15.7	15.7	15.7
	11	16.1	16.1	16.1	16.0
E	1	15.9	16.0	16.0	16.5
	2	16.1	16.3	16.0	16.0
	3	16.0	16.2	16.0	16.1
	4	16.3	16.5	16.4	16.4

Setting station	Regulator number	Testing station			
		1	2	3	4
J	1	16.2	16.1	15.8	16.0
	2	16.2	15.3	17.8	16.3
	3	16.4	16.7	16.5	16.5
	4	16.2	16.5	16.1	16.1
	5	16.1	16.4	16.1	16.3
	10	16.4	16.3	16.4	16.4
K	1	15.9	16.0	15.8	16.1
	2	15.8	15.7	16.7	16.0
	3	16.2	16.2	16.2	16.3
	4	16.2	16.3	15.9	16.3
	5	16.0	16.0	16.0	16.0
	6	16.0	16.4	16.2	16.2
	11	16.0	16.1	16.0	16.1

PROGRAM

```
  2  JOB       'Example S'
  3
  4  OPEN      'XS'; 2
  5
  6  "The data are presented as published in App. Stat., but with the setting
 -7  station explicitly given at the beginning of each row:
 -8  A  1 16.5 16.5 16.6 16.6
 -9  ...
-10  K 11 16.0 16.1 16.0 16.1 :
-11  To read the data set, structures of length 64 are needed:"
 12
 13  UNITS     [64]
 14  TEXT      NS; !T(A,B,C,D,E,F,G,H,J,K)
 15  FACTOR    [LEVELS=12] RGLTR
 16  FACTOR    [LABELS=NS] STTR
 17  READ      [CHANNEL=2] STTR,RGLTR,T[1...4]; FREP=LABELS,*,*,*
```

Identifier	Minimum	Mean	Maximum	Values	Missing
T[1]	15.40	16.03	16.90	64	0
T[2]	15.30	16.16	17.00	64	0
T[3]	15.30	16.13	17.80	64	0
T[4]	15.60	16.17	17.00	64	0

```
 18
 19  "The analysis is done on structures of length 256, with all the observations
-20  in a single variate VOLTAGE. This is most easily constructed by equating the
-21  four T variates to a matrix and transposing it:"
 22
 23  UNITS     [256]
 24  MATRIX    [ROWS=4; COLUMNS=64]  MT
 25  &         [ROWS=64; COLUMNS=4]  TM
 26  EQUATE    T; MT
 27  CALCULATE TM = TRANSPOSE(MT)
 28  EQUATE    TM; VOLTAGE
 29
 30  "There is no need to adjust VOLTAGE by subtracting 16, as in App. Stat.
-31  ANOVA begins by sweeping out the overall mean. If the mean is small, .01 or
-32  less, there may be some advantage in scaling the data before the analysis:
-33  the computations will be no more accurate but the output will be easier to
-34  comprehend with leading zeroes absent.
-35  (Similar remarks also apply to the Regression facilities.)
-36
-37  The factors are expanded, and one for testing stations defined:"
 38
 39  FACTOR    [LABELS=NS] SS; !(4(#STTR))
 40  FACTOR    [LEVELS=12] RG; !(4(#RGLTR))
 41  FACTOR    [LEVELS=4] TS; !((1...4)64)
 42
 43  "We set up a loop to get a simple analysis of variance within each setting
-44  station to obtain the residual mean squares required for Table S.2, App. Stat.
-45  The analysis of variance for setting station J omitting regulator 2 is also
-46  computed."
 47
 48  TREATMENT RG*TS
 49
 50  FOR       N=1...10
 51
 52   RESTRICT    VOLTAGE; SS.EQ.N
 53   PRINT       [IPRINT=*] 'Setting station no.',N; 4; 0
 54   ANOVA       [PRINT=AOV; FACTORIAL=1] VOLTAGE
 55
 56   IF N.EQ.9
 57    RESTRICT   VOLTAGE; SS.EQ.9 .AND. RG.NE.2
 58    PRINT      'Setting station no. 9 (J), Regulator 2 omitted'
 59    ANOVA      [PRINT=AOV; FACTORIAL=1] VOLTAGE
 60   ENDIF
 61
 62   RESTRICT    VOLTAGE
 63
 64  ENDFOR

Setting station no.   1
```

***** Analysis of variance *****

Variate: VOLTAGE

Source of variation	d.f.	s.s.	m.s.	v.r.
RG	7	2.74500	0.39214	12.29
TS	3	0.20000	0.06667	2.09
Residual	21	0.67000	0.03190	
Total	31	3.61500		

```
  65  "  .....    ( Only the output for the first setting is shown here.)
 -66
 -67  Now for the full analysis. Note that the restriction on VOLTAGE was removed
 -68  at the end of the FOR loop: this is good practice in general, and in this case
 -69  means that we can go straight into:"
  70
  71  TREATMENT SS*TS+SS/RG
  72  ANOVA     [PRINT=AOV; FACTORIAL=2]  VOLTAGE            "Table S.3 App. Stat."
```

***** Analysis of variance *****

Variate: VOLTAGE

Source of variation	d.f.	s.s.	m.s.	v.r.
SS	9	4.41905	0.49101	9.08
TS	3	0.78453	0.26151	4.84
SS.TS	27	0.90406	0.03348	0.62
SS.RG	54	9.48079	0.17557	3.25
Residual	162	8.76141	0.05408	
Total	255	24.34984		

```
  73  VCOMPONENTS  [FIXED=TS]  SS/RG                         "Table S.4 App. Stat."
  74  REML       [PRINT=COMPONENTS] VOLTAGE
```

*** Estimated Components of Variance ***

		s.e.
SS	0.01193	0.009060
SS.RG	0.03078	0.008474
units	0.05114	0.005261

```
  75  "Omitting the second regulator of setting station J:"
  76
  77  RESTRICT  VOLTAGE; SS.NE.9 .OR. RG.NE.2
  78  ANOVA     [PRINT=AOV; FACTORIAL=2]  VOLTAGE            "Table S.3 App. Stat."
```

***** Analysis of variance *****

Variate: VOLTAGE

Source of variation	d.f.	s.s.	m.s.	v.r.
SS	9	4.16209	0.46245	13.10
TS	3	0.89540	0.29847	8.45
SS.TS	27	0.72077	0.02670	0.76
SS.RG	53	9.42871	0.17790	5.04
Residual	159	5.61383	0.03531	
Total	251	20.82080		

```
  79  VCOMPONENTS  [FIXED=TS]  SS/RG                         "Table S.4 App. Stat."
  80  REML       [PRINT=COMPONENTS] VOLTAGE
```

*** Estimated Components of Variance ***

		s.e.
SS	0.01098	0.008732
SS.RG	0.03560	0.008599
units	0.03406	0.003532

```
81
82  STOP
```

******** End of Example S. Maximum of 159240 data units used at line 80 (21192 left)

SUGGESTED FURTHER WORK

1. Obtain and examine residuals to check the anomalous observations on J_2. Check for any other outliers.
2. Examine the assumption of normality.

Example T

Intervals between the failure of air-conditioning equipment in aircraft

DESCRIPTION OF DATA

The data in Table T.1, reported by Proschan (1963), are the intervals in service-hours between failures of the air-conditioning equipment in 10 Boeing 720 jet aircraft. It is required to describe concisely the variation within and between aircraft, with emphasis on the forms of the frequency distributions involved.

THE ANALYSIS

The data can be analysed in various ways to check consistency with an assumed theoretical model and to make comparisons between aircraft. Here we concentrate upon fitting by maximum likelihood a gamma distribution to the observed time intervals for each aircraft.

The gamma distribution of mean μ and index β,

$$\frac{(\beta/\mu)(\beta y/\mu)^{\beta-1}e^{-\beta y/\mu}}{\Gamma(\beta)} \qquad \text{(T.2, App. Stat.)}$$

reduces to the exponential distribution when $\beta = 1$, and comparisons between aircraft then become simply comparisons between means. Hence it is sensible not only to check on the consistency of the estimated βs but also to compare their common estimate with $\beta = 1$. Instructions are given for fitting:

1. separate gamma distributions to all aircraft, i.e. parameters μ_i, β_i for $i = 1, \ldots, 10$;
2. separate gamma distributions but with common β;
3. common gamma distribution to all aircraft, with 2 parameters;
4. separate exponential distributions to all aircraft ($\beta = 1$, separate μ_i);
5. common exponential distribution to all aircraft ($\beta = 1$).

Table T.1 Intervals between failures (operating hours)

				Aircraft number					
1	2	3	4	5	6	7	8	9	10
413	90	74	55	23	97	50	359	487	102
14	10	57	320	261	51	44	9	18	209
58	60	48	56	87	11	102	12	100	14
37	186	29	104	7	4	72	270	7	57
100	61	502	220	120	141	22	603	98	54
65	49	12	239	14	18	39	3	5	32
9	14	70	47	62	142	3	104	85	67
169	24	21	246	47	68	15	2	91	59
447	56	29	176	225	77	197	438	43	134
184	20	386	182	71	80	188		230	152
36	79	59	33	246	1	79		3	27
201	84	27	15	21	16	88		130	14
118	44	153	104	42	106	46			230
34	59	26	35	20	206	5			66
31	29	326		5	82	5			61
18	118			12	54	36			34
18	25			120	31	22			
67	156			11	216	139			
57	310			3	46	210			
62	76			14	111	97			
7	26			71	39	30			
22	44			11	63	23			
34	23			14	18	13			
	62			11	191	14			
	130			16	18				
	208			90	163				
	70			1	24				
	101			16					
	208			52					
				95					

The FITNONLINEAR command is used to obtain maximum likelihood estimates. The estimate of μ_i is equal to $\bar{y}_i$, the sample mean; this is used as the initial approximation to increase the efficiency of iteration for estimation of β_i, or β.

The maximised log-likelihood is computed for the fitted gamma distributions and for the fitted exponential distributions. The hypothesis of a common distribution for all aircraft, or of gamma versus exponential distributions, can be tested by direct comparison of the appropriate maximised log-likelihoods. The conclusions are not clear-cut and are discussed in *App. Stat.*

A line-printer plot shows the observations for two of the aircraft plotted against exponential order statistics.

PROGRAM

```
  2  JOB       'Example T'
  3
  4  OPEN      'XT'; 2
  5
  6  "Data:
 -7  413 14 58 37 100 65 9 169 447 184 36 201 118 34 31 18 18 67 57 62 7 22 34:
 -8  ...
 -9  102 209 14 57 54 32 67 59 134 152 27 14 230 66 61 34:
-10
-11  The data for this example are variates of differing lengths. The values of
-12  each should be terminated by an end-of-data marker (:) and two options of
-13  READ used: SERIAL=YES to tell READ that the structure values will appear
-14  serially, and SETNVALUES=YES to note the numbers of values read for each
-15  variate:"
 16
 17  READ      [CHANNEL=2; SERIAL=YES; SETNVALUES=Y] AIRCRAFT[1...10]
```

Identifier	Minimum	Mean	Maximum	Values	Missing	
AIRCRAFT[1]	7.00	95.70	447.00	23	0	Skew
AIRCRAFT[2]	10.00	83.52	310.00	29	0	Skew
AIRCRAFT[3]	12.0	121.3	502.0	15	0	Skew
AIRCRAFT[4]	15.0	130.9	320.0	14	0	
AIRCRAFT[5]	1.00	59.60	261.00	30	0	Skew
AIRCRAFT[6]	1.00	76.81	216.00	27	0	
AIRCRAFT[7]	3.00	64.13	210.00	24	0	
AIRCRAFT[8]	2.0	200.0	603.0	9	0	
AIRCRAFT[9]	3.0	108.1	487.0	12	0	Skew
AIRCRAFT[10]	14.00	82.00	230.00	16	0	

```
 18
 19  "If we did not need to estimate the dispersion as an explicit parameter we
-20  could use an option of MODEL to set DISTRIBUTION=GAMMA. For this example we
-21  use the FITNONLINEAR command which fits models containing linear and nonlinear
-22  terms and also finds the minima of functions. A MODEL directive specifies a
-23  scalar which will contain the value of the function.
-24
-25  LLGAMMA(x;m;b) evaluates the log-likelihood of the gamma distribution, (apart
-26  from a term SUM(LOG(x)), given the values of its parameters. (Since we wish to
-27  maximise the likelihood, the negative of the log-likelihood is calculated.)
-28
-29  RCYCLE notes the names of the parameters. Initial estimates may be supplied,
-30  to shorten the process; since the sample mean is the estimate of the first
-31  parameter, it is used.
-32
-33      (i) Separate Gamma distributions
-34
-35  The fits are obtained in a FOR loop over the 10 aircraft, and the estimates
-36  saved (in variates of length 2) for display later. At the same time,
-37  the quantity Ni*LOG(Yi/Ni) is formed, and the maximum log-likelihoods for
-38  the gamma (LLG20) and exponential (LLE10) distributions calculated."
 39
 40  VARIATE   [NVALUES=2] AC[1...10]
 41  SCALAR    LLE10,LLG20,SE,SL; 0
 42
 43  FOR       A=AIRCRAFT[]; P=AC[]; J=1...10
 44
 45   PRINT        [IPRINT=*] 'Aircraft no.',J; FIELD=12,3; DECIMALS=0
 46   CALCULATE    MI = MEAN(A)
 47   MODEL        [FUNCTION=FG]
 48   EXPRESSION   EG; !E(FG=-LLGAMMA(A; Mu; Beta))
```

```
49    RCYCLE         Mu,Beta; INITIAL=MI,*; LOWER=*,0.0005
50    FITNONLINEAR [CALCULATE=EG]
51    RKEEP          ESTIMATES = P
52    CALCULATE      SE = SE+( L= ( N = NVAL(A) )*LOG(MI) )  & LLE10 = LLE10-L-N
53    &              SL = SL+SUM(LOG(A))  & LLG20 = LLG20-FG
54
55    ENDFOR
```

Aircraft no. 1

***** Results of optimization *****

*** Minimum function value: ***

        36.8027

*** Estimates of parameters ***
                           sq. root of
                  estimate   2nd derivs
Mu   95.8   28.8
Beta   0.966   0.353

```
56    "  ...... (Only the output for the first aircraft is shown.)
-57
-58   Summarising all the estimates:"
59
60    PRINT      !T(Mu,Beta),AC[]; FIELD=5,10(7); DECIMALS=2    "Table T.2 App. Stat."
```

```
        AC[1]  AC[2]  AC[3]  AC[4]  AC[5]  AC[6]  AC[7]  AC[8]  AC[9] AC[10]
   Mu   95.79  83.52 121.29 130.86  59.61  76.82  64.13 200.06 108.09  82.00
 Beta    0.97   1.67   0.91   1.61   0.81   1.13   1.06   0.46   0.71   1.75
```

```
61    CALCULATE      LLG20=LLG20-SL
62
63    "(ii) Separate means for each aircraft with common Beta
-64
-65   Pointers are declared to hold the means, estimates and contributions to
-66   log-likelihood for each aircraft."
67
68    POINTER        [NVALUES=10] G,M,MU
69    SCALAR         G[],M[],MU[]
70    CALCULATE      M[] = MEAN(AIRCRAFT[])
71    MODEL          [FUNCTION=F]
72    EXPRESSION     E[1]; !E(G[]=-(LLGAMMA(AIRCRAFT[]; MU[]; Beta)))
73    &              E[2]; !E(F=G[1]+G[2]+G[3]+G[4]+G[5]+G[6]+G[7]+G[8]+G[9]+G[10])
74    RCYCLE         MU[],Beta; INITIAL=M[],*; LOWER=0.0005
75    FITNONLINEAR [CALCULATE=E]
```


***** Results of optimization *****

*** Minimum function value: ***

        313.052

*** Estimates of parameters ***
                           sq. root of
                  estimate   2nd derivs
MU[1]   95.7   28.1
MU[2]   83.5   21.9
MU[3]   121.3   44.2
MU[4]   130.9   49.3
MU[5]   59.6   15.3
MU[6]   76.8   20.8
MU[7]   64.1   18.5

```
MU[8]                 200.0        94.0
MU[9]                 108.1        44.0
MU[10]                 82.0        28.9
Beta                   1.006        0.125
```

```
 76   "Note that aircraft no. 8 has a low value of beta, 0.46. If this aircraft is
-77   omitted the common estimate of beta is 1.07 as in App. Stat."
 78   CALCULATE   LLG11 = -F-SL
 79
 80   "(iii) A single distribution over all aircraft
-81
-82   The values for each aircraft are combined into one variate."
 83
 84   VARIATE    ALL; !(#AIRCRAFT[])
 85   SCALAR     MM,MB
 86   CALCULATE MP = VMEAN(AC)  & MM,MB = MP$[1,2]
 87   MODEL      [FUNCTION=F]
 88   EXPRESSION EA; !E(F=-LLGAMMA(ALL; Mu; Beta))
 89   RCYCLE     Mu,Beta; INITIAL=MM,MB; LOWER=*,0.0005
 90   FITNONLINEAR [CALCULATE=EA]
```

```
***** Results of optimization *****

*** Minimum function value: ***

      322.602
```

```
*** Estimates of parameters ***
                              sq. root of
                 estimate     2nd derivs
Mu                90.89          9.44
Beta               0.935         0.111
```

```
 91   CALCULATE LLG2 = -F-SL
 92
 93   "We can now display the maximised log-likelihoods:"
 94
 95   PRINT      LLG20,LLG11,LLG2,LLE10; DECIMALS=2

       LLG20         LLG11          LLG2          LLE10
     -1078.68      -1086.53      -1096.08      -1086.54
```

```
 96
 97   "Twice the difference in the maximimum log-likelihoods for models with
-98   separate and common indices are now computed; these differnces are
-99   approximately chi-sq. with 9 d.f (T.1 App. Stat.)."
100
101   CALCULATE CHISQE = 2*(NVAL(ALL)*LOG(MEAN(ALL))-SE)
102    &        CHISQG = 2*(LLG20-LLG11)
103   PRINT     CHISQE,CHISQG

      CHISQE       CHISQG
      19.70        15.70
```

```
104   "(The value for CHISQG given in App. Stat. is incorrect.)
-105
-106  Finally, the ordered observations for aircraft 8 and 9 are compared
-107  with the exponential order statistics (derived by CALCULATE)."
108
109   FOR        A=AIRCRAFT[8,9]; B = EOS[8,9]
110    SORT      A
111    CALCULATE N = NVAL(A)
112    VARIATE   [NVALUES=N]  B
113    CALCULATE B$[1] = 1/N
114    FOR       J = 2...N
115     CALCULATE K = J-1
116      &        B$[J] = B$[K]+1/(N-K)
117    ENDFOR
```

```
118   ENDFOR
119
120   GRAPH    [YTITLE='Times between failures';                              \
121            XTITLE='exponential order statistics'; NROWS=22]               \
122            AIRCRAFT[8,9]; EOS[8,9]; SYMBOLS='8','9'        "Fig T.1 App. Stat."
```

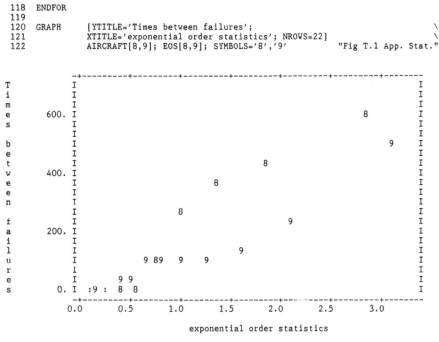

exponential order statistics

```
123   "The symbol ':' indicates coincident points."
124
125   STOP
```

******** End of Example T. Maximum of 16846 data units used at line 118 (33144 left)

SUGGESTED FURTHER WORK

Carry out a similar analysis using the Weibull distribution and compare
with the above results.

Example U

Survival times of leukemia patients

DESCRIPTION OF DATA

The data in Table U.1 from Feigl and Zelen (1965) are time to death, Y, in weeks from diagnosis and $\log_{10}$(initial white blood cell count), x, for 17 patients suffering from leukemia. The relation between Y and x is the main aspect of interest.

Table U.1 Survival time Y in weeks and $\log_{10}$(initial white blood cell count), x, for 17 leukemia patients

x	Y	x	Y	x	Y
3.36	65	4.00	121	4.54	22
2.88	156	4.23	4	5.00	1
3.63	100	3.73	39	5.00	1
3.41	134	3.85	143	4.72	5
3.78	16	3.97	56	5.00	65
4.02	108	4.51	26		

THE ANALYSIS

General considerations of survival times frequently suggest an exponential distribution. In *App. Stat.* it is assumed that

$$Y_i = \beta_0 \exp\{\beta_1(x_i - \bar{x})\}\varepsilon_i \qquad \text{(U.1, } App. \ Stat.)$$

where ε_i is a random variable which is exponentially distributed with unit mean. Writing the model in this form, i.e. deviations $(x_i - \bar{x})$ rather than simply x_i, leads to zero covariance between the maximum likelihood estimates $\hat{\beta}_0$ and $\hat{\beta}_1$ which is helpful if determining a standard error for predicted values of Y and which with more complex sets of data would have numerical analytical advantages.

The above model is a generalised linear model for which

$$E(Y_i) = \exp \{ \log \beta_0 + \beta_1(x_i - \bar{x}) \}$$

and this is conveniently fitted by taking a log link function with a gamma distribution (setting the dispersion option equal to unity).

High-quality graphics are used to illustrate the fitted model; see Fig. U.1 after the Genstat program.

PROGRAM

```
  2  JOB        'Example U'
  3
  4  OPEN       'XU'; 2
  5
  6  "Data:
 -7  3.36  65    4.00 121    4.54  22
 -8  ...
 -9  4.02 108    4.51  26 : "
 10
 11  UNITS      [17]
 12  READ       [CHANNEL=2]  X,Y
```

Identifier	Minimum	Mean	Maximum	Values	Missing
X	2.880	4.096	5.000	17	0
Y	1.00	62.47	156.00	17	0

```
 13
 14  "Deviations of X from its mean are much used subsequently, so:"
 15
 16  CALCULATE XD = X-MEAN(X)
 17
 18  "The generalized linear model E(Y) = EXP(LOG(b0)+b1*XD), with a
-19  log link function and gamma distribution, is fitted."
 20
 21  MODEL      [DISTRIBUTION=GAMMA; LINK=LOG; DISPERSION=1] Y
 22
 23  "The dispersion option is set to 1, the constant of proportionality for the
-24  gamma distribution, in order to fit the exponential model of App. Stat.
-25  The deviance is -2*loglikelihood of the exponential distribution."
 26
 27  FIT        [PRINT=MODEL,SUMMARY,ESTIMATES,FITTED] XD
```

***** Regression Analysis *****

Response variate: Y
    Distribution: Gamma
    Link function: Log
    Fitted terms: Constant, XD

*** Summary of analysis ***
    Dispersion parameter is 1

	d.f.	deviance	mean deviance	deviance ratio
Regression	1	6.83	6.826	5.26
Residual	15	19.46	1.297	
Total	16	26.28	1.643	

* MESSAGE: The following units have large standardized residuals:
            6   -2.21
            9   -2.21

* MESSAGE: The following units have high leverage:
            4   0.29

*** Estimates of regression coefficients ***

	estimate	s.e.	t
Constant	3.934	0.243	16.22
XD	-1.109	0.400	-2.78

* MESSAGE: s.e.s are based on dispersion parameter with value 1

*** Fitted values and residuals ***

Unit	Response	Fitted value	Standardized residual	Leverage
1	65.00	115.60	-0.57	0.15
2	121.00	56.84	0.89	0.06
3	22.00	31.23	-0.35	0.09
4	156.00	196.85	-0.27	0.29
5	4.00	44.04	-1.78	0.06
6	1.00	18.75	-2.21	0.19
7	100.00	85.68	0.17	0.09
8	39.00	76.69	-0.63	0.08
9	1.00	18.75	-2.21	0.19
10	134.00	109.36	0.23	0.13
11	143.00	67.13	0.90	0.07
12	5.00	25.58	-1.37	0.12
13	16.00	72.55	-1.26	0.07
14	56.00	58.77	-0.05	0.06
15	65.00	18.75	1.74	0.19
16	108.00	55.60	0.77	0.06
17	26.00	32.29	-0.22	0.09
Mean		62.47	63.79 -0.37	0.12

```
  28      "The residual deviance, 19.46, is asymptotically distributed as a chi-squared
 -29      variable with 15 d.f. if the model assumptions are true."
  30
  31 RKEEP       DEVIANCE=CHI; DF=DF; ESTIMATES=E
  32 CALCULATE PROB = 1-CHISQ(CHI;DF)
  33 PRINT       CHI,DF,PROB
```

CHI	DF	PROB
19.46	15.00	0.1938

```
  34 SCALAR      b1
  35 CALCULATE b1 = E$[2]
  36   &         b0 = MEAN(Y*EXP(-b1*XD))
  37   &         s0 = b0/SQRT(NVAL(XD))
  38   &         s1 = 1/SQRT(SUM(XD*XD))
  39   &         Fitted = b0*EXP(b1*XD)
  40 PRINT       b0,s0,b1,s1
```

b0	s0	b1	s1
51.11	12.40	-1.109	0.3997

```
  41 OPEN        'U.PLT'; 6; Graphics
  42 DEVICE      6
  43 AXES        WINDOW=1; XTITLE='Log(initial cell count)';                              \
  44             YTITLE='Survival time (weeks)'; YLOWER=0; YMARKS=!(50,100,150);          \
  45             XLOWER=2.5; XMARKS=!(3.0,3.5...5.0)
  46 FRAME       WINDOW=1,2; YLOWER=0.0,0.4; YUPPER=0.7,0.6;XLOWER=0.0,0.4; XUPPER=0.7
  47 PEN         1,2; METHOD=LINE,POINT; LINESTYLE=1; SYMBOLS = 0,1
  48 DGRAPH      Fitted,Y; X; PEN=1,2;DESCRIPTION = 'Exponential model',*
  49
  50 STOP
```

******** End of Example U. Maximum of 14850 data units used at line 39 (35140 left)

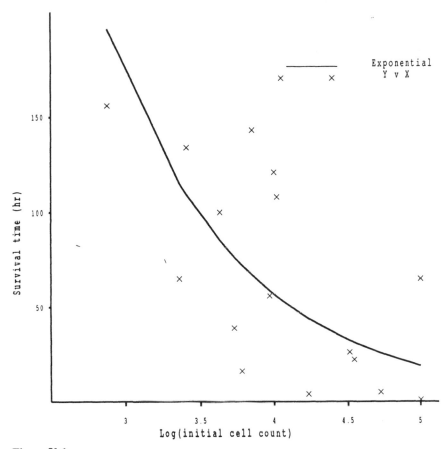

Figure U.1

SUGGESTED FURTHER WORK

1. Fit the model in the nonorthogonal form, i.e. $E(Y_i) = \gamma_0 \exp(\gamma_1 x_i)$, and compare with preceding analysis.
2. Determine a confidence interval for the expected value of Y for a given value x.
3. Consider what other forms of model might be appropriate.

Example V

A retrospective study with binary data

DESCRIPTION OF DATA

In a retrospective study of the possible effect of blood group on the incidence of peptic ulcers, Woolf (1955) obtained data from three cities. Table V.1 gives for each city data for blood groups O and A only. In each city, blood group is recorded for peptic-ulcer subjects and for a control series of individuals not having peptic ulcers.

Table V.1 Blood groups for peptic ulcer and control subjects

	Peptic ulcer		Control	
	Group O	Group A	Group O	Group A
London	911	579	4578	4219
Manchester	361	246	4532	3775
Newcastle	396	219	6598	5261

THE ANALYSIS

As discussed in *App. Stat.*, we wish to study how the probability of a peptic ulcer depends on blood group, but the data have been collected in an inverse way from samples of peptic ulcer and non-peptic ulcer subjects. However, the log of the odds ratio is given by

$$\Delta = \log\left\{\frac{\text{pr (ulcer|A)}}{\text{pr (ulcer|O)}} \times \frac{\text{pr (no ulcer|O)}}{\text{pr (no ulcer|A)}}\right\} \quad (V.1, \text{\textit{App. Stat.}})$$

$$= \log\left\{\frac{\text{pr (A|ulcer)}}{\text{pr (O|ulcer)}}\right\} - \log\left\{\frac{\text{pr (A|no ulcer)}}{\text{pr (O|no ulcer)}}\right\}.$$

We could estimate Δ by logistic regression, fitting the model

$$\log\left\{\frac{\text{pr(A)}}{\text{pr(O)}}\right\} = \begin{cases} \mu, & \text{non-ulcer group} \\ \mu + \Delta, & \text{ulcer group.} \end{cases}$$

as shown in the programs for Examples H, M and X.

Here instead we demonstrate the use of straightforward CALCU-LATE commands to follows the approach in *App. Stat.* based upon large-sample approximations. The estimate of Δ_j, the parameter for city j $(j = 1, 2, 3)$ is

$$\tilde{\Delta}_j = \log\left(\frac{R_{2j}}{n_{2j} - R_{2j}}\right) - \log\left(\frac{R_{1j}}{n_{1j} - R_{1j}}\right) \qquad (\text{V.4, } App.\ Stat.)$$

and this has large-sample variance

$$v_j = \frac{1}{R_{2j}} + \frac{1}{n_{2j} - R_{2j}} + \frac{1}{R_{1j}} + \frac{1}{n_{1j} - R_{1j}}, \qquad (\text{V.5, } App.\ Stat.)$$

assuming the Rs to be independently bionomially distributed.

Thus a χ^2 statistic for conformity with constant Δ_j is

$$\chi^2 = \Sigma(\tilde{\Delta}_j - \tilde{\Delta}.)^2/v_j$$
$$= \Sigma\tilde{\Delta}_j^2/v_j - (\Sigma\tilde{\Delta}_j/v_j)^2(\Sigma 1/v_j)^{-1}, \qquad (\text{V.6, } App.\ Stat.)$$

where

$$\tilde{\Delta}. = \frac{\Sigma\tilde{\Delta}_j/v_j}{\Sigma 1/v_j} \qquad (\text{V.7, } App.\ Stat.)$$

is the weighted mean with large-sample variance

$$(\Sigma 1/v_j)^{-1}. \qquad (\text{V.8, } App.\ Stat.)$$

If in fact all Δ_j are equal, the large-sample distribution of (V.6, *App. Stat.*) is χ^2 with two degrees of freedom.

PROGRAM

```
 2  JOB        'Example V'
 3
 4  OPEN       'XV'; 2
 5
 6  "Data:
-7     911  579 4578 4219
-8     361  246 4532 3775
-9     396  219 6598 5261 : "
10
11  UNITS      [3]
12  TEXT       CITY; !T(London,Manchester,Liverpool)
13  READ       [CHANNEL=2] NR2,R2,NR1,R1
```

```
       Identifier    Minimum       Mean    Maximum     Values    Missing
              NR2       361.0      556.0      911.0          3          0
               R2       219.0      348.0      579.0          3          0
              NR1        4532       5236       6598          3          0
               R1        3775       4418       5261          3          0
14
15   "This example only needs simple CALCULATE statements."
16
17   CALCULATE  DELTA = LOG(R2*NR1/(NR2*R1))
18      &       LSV = 1/R2+1/NR2+1/R1+1/NR1
19      &       LSE = SQRT(LSV)
20      &       EXP = EXP(DELTA)
21   PRINT      CITY,DELTA,LSE,EXP; FIELD=15; DECIMALS=0,4,5,3
```

CITY	DELTA	LSE	EXP
London	-0.3716	0.05727	0.690
Manchester	-0.2008	0.08556	0.818
Liverpool	-0.3659	0.08622	0.694

```
22
23   CALCULATE  WEIGHTS = 1/LSV  & W = SUM(WEIGHTS)
24      &       MNDLT = SUM(WEIGHTS*DELTA)/W
25      &       ESEMN = SQRT(1/W)
26      &       EXPMN = EXP(MNDLT)
27      &       CHISQ = SUM(WEIGHTS*(DELTA-MNDLT)**2)
28
29   PRINT      'Pooled',MNDLT,ESEMN,EXPMN,CHISQ; FIELD=15; DECIMALS=0,4,5,3,2
```

	MNDLT	ESEMN	EXPMN	CHISQ
Pooled	-0.3297	0.04167	0.719	2.98

```
30
31   TEXT       CITY; !T(#CITY,'Pooled')
32   VARIATE    DELTA; !(#DELTA,#MNDLT)
33      &       LSE; !(#LSE,#ESEMN)
34      &       EXP; !(#EXP,#EXPMN)
35   PRINT      CITY,DELTA,LSE,EXP; FIELD=15; DECIMALS=0,4,5,3  "Table V.2 App. Stat."
```

CITY	DELTA	LSE	EXP
London	-0.3716	0.05727	0.690
Manchester	-0.2008	0.08556	0.818
Liverpool	-0.3659	0.08622	0.694
Pooled	-0.3297	0.04167	0.719

```
36
37   STOP
```

******** End of Example V. Maximum of 6774 data units used at line 35 (42586 left)

SUGGESTED FURTHER WORK

Fit a logistic regression

1. separately to each city to estimate Δ_j ($j = 1, 2, 3$);
2. across all cities to estimate a common Δ.

Compare the estimates with those obtained above. Check for conformity with a common Δ by differencing the maximised log-likelihoods (deviances).

Example W

Housing and associated factors

DESCRIPTION OF DATA

The data in Table W.1 (Madsen, 1976) relate to an investigation into satisfaction with housing conditions in Copenhagen. A total of 1681 residents from selected areas living in rented homes built between 1960 and 1968 were questioned on their satisfaction, the degree of contact with other residents and their feeling of influence on apartment management. The purpose of the investigation was to study association between these three factors and the type of housing.

Table W.1 1681 persons classified according to satisfaction, contact, influence and type of housing

Contact		Low			High		
Satisfaction		Low	Medium	High	Low	Medium	High
Housing	*Influence*						
Tower blocks	Low	21	21	28	14	19	37
	Medium	34	22	36	17	23	40
	High	10	11	36	3	5	23
Apartments	Low	61	23	17	78	46	43
	Medium	43	35	40	48	45	86
	High	26	18	54	15	25	62
Atrium houses	Low	13	9	10	20	23	20
	Medium	8	8	12	10	22	24
	High	6	7	9	7	10	21
Terraced houses	Low	18	6	7	57	23	13
	Medium	15	13	13	31	21	13
	High	7	5	11	5	6	13

THE ANALYSIS

Type of housing is treated as an explanatory variable, with satisfaction, contact and influence being treated as response variables.

We first inspect the distribution for each response variable within each type of housing. There are marked differences in distribution between types of housing. Hence we fit loglinear models in the three response variables, separately within each type of housing, but hoping to obtain reasonably consistent conclusions from the separate models.

The data for tower blocks, atrium houses and terraced houses can be explained by a loglinear model containing main effects plus the interaction $S \times I$ (satisfaction $\times$ influence). Apartments appear to need all three two-factor interactions, although for a simplified interpretation we take the model containing just the one interaction. Interpretation is based upon the ratios of fitted frequencies under the model containing $S \times I$ to those under an independence model; see *App. Stat.* for further details.

PROGRAM

```
  3   JOB        'Example W'
  4
  5   OPEN       'XW'; 2
  6
  7   "Data:
 -8    21 21 28 14 19 37
 -9    ...
-10     7  5 11  5  6 13 : "
 11
 12   UNITS      [72]
 13   TEXT       NC; !T(Low,High)
 14    &         NX; !T(Low,Medium,High)
 15    &         NH; !T('Tower blocks','Apartments','Atrium houses','Terraced Houses')
 16   FACTOR     [LABEL=NC] Contact
 17    &         [LABEL=NX] Stsfctn,Fluence
 18    &         [LABEL=NH] Housing
 19   TABLE      [CLASS=Housing,Fluence,Contact,Stsfctn] Counts
 20   READ       [CHANNEL=2] Counts
```

```
      Identifier   Minimum     Mean    Maximum    Values   Missing
          Counts      3.00    23.35      86.00        72         0    Skew
 21
 22   "Two way tables are defined to extract the marginal totals required to
-23   form the percentages:"
 24
 25   TABLE      [CLASS=Housing,Stsfctn]  CHS,%HS
 26    &         [CLASS=Housing,Contact]  CHC,%HC
 27    &         [CLASS=Housing,Fluence]  CHI,%HI
 28    &         [CLASS=Housing]          TOTALS
 29   CALCULATE  CHS,CHC,CHI,TOTALS = Counts
 30    &         %HS,%HC,%HI = 100*CHS,CHC,CHI/TOTALS
 31   PRINT      %HS,%HC,%HI,TOTALS; FIELD=10; DECIMALS=1,1,1,0  "Table W.2 App. Stat."
```

```
                     %HS
       Stsfctn      Low     Medium      High
       Housing
  Tower blocks     24.7       25.2      50.0
    Apartments     35.4       25.1      39.5
```

Atrium houses	26.8	33.1	40.2
Terraced Houses	48.0	26.7	25.3

	%HC	
Contact	Low	High
Housing		
Tower blocks	54.7	45.2
Apartments	41.4	58.6
Atrium houses	34.3	65.7
Terraced Houses	34.3	65.7

	%HI		
Fluence	Low	Medium	High
Housing			
Tower blocks	35.0	43.0	22.0
Apartments	35.0	38.8	26.1
Atrium houses	39.7	35.1	25.1
Terraced Houses	44.8	38.3	17.0

	TOTALS
Housing	
Tower blocks	400
Apartments	765
Atrium houses	239
Terraced Houses	277

```
32   "So far, it has been natural and convenient to store the information in tables.
-33   To fit models, the data are transferred to a vector and factor values generated.
-34   All the real work is done in a loop over the four housing types. Normally, at
-35   least summary information would be output, but here all printing is suppressed
-36   and the values required for display are stored in various structures by RKEEP
-37   and CALCULATE."
38
39   VARIATE    VCOUNT; !(#Counts)
40   GENERATE   Housing,Fluence,Contact,Stsfctn
41   POINTER    DV; !P(Towers,Aparts,Atrium,Terraced)
42   VARIATE    F[1...4]  & [NVAL=4] df,DV[]
43   TABLE      [CLASS=Fluence,Stsfctn] T1[1...4],T2[1...4]
44   MODEL      [DISTRIBUTION=POISSON; LINK=LOGARITHM] VCOUNT
45
46   FOR        H=1...4
47
48    RESTRICT  VCOUNT; Housing.EQ.H
49    TERMS     Fluence*Contact*Stsfctn
50
51    FIT       [PRINT=*]  Fluence+Contact+Stsfctn
52    RKEEP     DEVIANCE=S1; DF=S2; FITTED=F2[H]
53    CALCULATE DV[H]$[1],df$[1] = S1,S2
54    PREDICT   [PRINT=*; PREDICTIONS=T1[H]] Fluence,Stsfctn
55
56    FIT       [PRINT=*]  Fluence+Contact+Stsfctn+Contact.Stsfctn
57    RKEEP     DEVIANCE=S1; DF=S2
58    CALCULATE DV[H]$[2],df$[2] = S1,S2
59
60    FIT       [PRINT=*]  Fluence+Contact+Stsfctn+Contact.Fluence
61    RKEEP     DEVIANCE=S1; DF=S2
62    CALCULATE DV[H]$[3],df$[3] = S1,S2
63
64    FIT       [PRINT=*]  Fluence+Contact+Stsfctn+Stsfctn.Fluence
65    RKEEP     DEVIANCE=S1; DF=S2; FITTED=F1[H]
66    CALCULATE DV[H]$[4],df$[4] = S1,S2
67    PREDICT   [PRINT=*; PREDICTIONS=T2[H]] Fluence,Stsfctn
68
69    RESTRICT  VCOUNT
70
71   ENDFOR
72
73   "We print a table of deviances for the fitted models (compared with the full
-74   model). An unnamed text structure is supplied to label the rows:"
75
76   PRINT    !T('(a) Main effects            ','(b) Main effects plus C x S',    \
```

```
77                          '              C x I',' 
78             df,DV[]; FIELD=30,7,4(10); DECIMALS=0,0,4(1)
```
```
                                S x I'),        \
                    "Table W.3 App. Stat."
```

	df	Towers	Aparts	Atrium	Terraced
(a) Main effects	12	28.8	98.0	11.6	40.4
(b) Main effects plus C x S	10	22.1	90.2	9.1	36.2
C x I	10	23.8	91.4	11.4	30.2
S x I	8	14.3	22.4	3.9	14.2

```
79  TABLE      [CLASS=Housing,Fluence,Contact,Stsfctn] Residual,Fit1,Fit2
80  CALCULATE  V1,V2 = VSUM(F1,F2)
81  EQUATE     V1,V2; Fit1,Fit2
82  PRINT      [NDOWN=3; INTERLEAVE=3] Counts,Fit1,Fit2;FIELD=10; DECIMALS=0,1,1
83  "Table W.5 App. Stat. is printed separately, sideways, on the next page."
84
85  "Observed counts and fitted frequencies for the model including the interaction
-86  of satisfaction and influence (Fit1) and the independence model (Fit2)."
87
88  CALCULATE  Residual = (Counts-Fit1)/SQRT(Fit1)
89  PRINT      [NDOWN=2] Residual; FIELD=9; DECIMALS=1          "Table W.4 App. Stat."
```

		Residual					
	Contact	Low			High		
	Stsfctn	Low	Medium	High	Low	Medium	High
Housing	Fluence						
Tower blocks	Low	0.4	-0.2	-1.3	-0.5	0.2	1.4
	Medium	1.2	-0.5	-0.9	-1.3	0.6	1.0
	High	1.1	0.8	0.7	-1.2	-0.8	-0.7
Apartments	Low	0.4	-1.0	-1.6	-0.4	0.9	1.3
	Medium	0.9	0.3	-1.7	-0.7	-0.3	1.4
	High	2.2	0.0	0.9	-1.8	0.0	-0.7
Atrium houses	Low	0.5	-0.6	-0.1	-0.4	0.4	0.1
	Medium	0.7	-0.7	-0.1	-0.5	0.5	0.1
	High	0.7	0.5	-0.4	-0.5	-0.3	0.3
Terraced Houses	Low	-1.5	-1.3	0.1	1.1	0.9	0.0
	Medium	-0.2	0.4	1.4	0.1	-0.3	-1.0
	High	1.4	0.6	1.0	-1.0	-0.5	-0.7

```
90  TABLE      [CLASS=Housing,Fluence,Stsfctn] t1,t2,Ratios
91  EQUATE     T1,T2; t1,t2
92  CALCULATE  Ratios = t2/t1
93  PRINT      Ratios; FIELD=10; DECIMALS=2                     "Table W.6 App. Stat."
```

		Ratios		
	Stsfctn	Low	Medium	High
Housing	Fluence			
Tower blocks	Low	1.01	1.13	0.93
	Medium	1.20	1.04	0.88
	High	0.60	0.72	1.34
Apartments	Low	1.46	1.03	0.57
	Medium	0.86	1.07	1.07
	High	0.58	0.86	1.47
Atrium houses	Low	1.30	1.02	0.79
	Medium	0.80	1.08	1.07
	High	0.81	0.86	1.24
Terraced Houses	Low	1.26	0.88	0.64
	Medium	0.90	1.20	0.97
	High	0.53	0.88	2.02

```
94  STOP
```

******** End of Example W. Maximum of 18552 data units used at line 71 (31438 left)

SUGGESTED FURTHER WORK

Consider ways in which the order of the categories low, medium and high for satisfaction and influence may be taken into account.

Table V.5, App. Stat.

Housing	Fluence	Contact Stsfctn	Low Low	Medium	High	High Low	Medium	High
Tower blocks	Low	Counts	21	21	28	14	19	37
		Fit1	19.2	21.9	35.6	15.8	18.1	29.4
		Fit2	19.0	19.4	38.3	15.7	16.0	31.7
	Medium	Counts	34	22	36	17	23	40
		Fit1	27.9	24.6	41.6	23.1	20.4	34.4
		Fit2	23.3	23.8	47.1	19.3	19.7	38.9
	High	Counts	10	11	36	3	5	23
		Fit1	7.1	8.8	32.3	5.9	7.2	26.7
		Fit2	11.9	12.2	24.1	9.9	10.1	19.9
Apartments	Low	Counts	61	23	17	78	46	43
		Fit1	57.6	28.6	24.9	81.4	40.4	35.1
		Fit2	39.3	27.9	43.8	55.6	39.4	62.0
	Medium	Counts	43	35	40	48	45	86
		Fit1	37.7	33.2	52.2	53.3	46.8	73.8
		Fit2	43.6	30.9	48.6	61.6	43.7	68.7
	High	Counts	26	18	54	15	25	62
		Fit1	17.0	17.8	48.1	24.0	25.2	67.9
		Fit2	29.4	20.8	32.7	41.5	29.4	46.2
Atrium houses	Low	Counts	13	9	12	20	23	20
		Fit1	11.3	11.0	10.3	21.7	21.0	19.7
		Fit2	8.7	10.8	13.1	16.7	20.6	25.1
	Medium	Counts	8	8	12	10	22	24
		Fit1	6.2	10.3	12.4	11.8	19.7	23.6
		Fit2	7.7	9.5	11.6	14.8	18.2	22.2
	High	Counts	6	7	9	7	10	21
		Fit1	4.5	5.8	10.3	8.5	11.2	19.7
		Fit2	5.5	6.8	8.3	10.6	13.0	15.8
Terraced Houses	Low	Counts	18	6	7	57	23	13
		Fit1	25.7	9.9	6.9	49.3	19.1	13.1
		Fit2	20.4	11.4	20.7	39.1	21.8	20.6
	Medium	Counts	15	13	13	31	21	13
		Fit1	15.8	11.7	8.9	30.2	22.3	17.1
		Fit2	17.5	9.7	9.2	33.4	18.6	17.6
	High	Counts	7	5	11	5	6	13
		Fit1	4.1	3.8	8.2	7.9	7.2	15.8
		Fit2	7.7	4.3	4.1	14.8	8.2	7.8

Example X

Educational plans of Wisconsin schoolboys

DESCRIPTION OF DATA

Sewell and Shah (1968) have investigated for some Wisconsin highschool 'senior' boys and girls the relationship between variables:

1. socioeconomic status (high, upper middle, lower middle, low);
2. intelligence (high, upper middle, lower middle, low);
3. parental encouragement (low, high);
4. plans for attending college (yes, no).

The data for boys are given in Table X.1.

THE ANALYSIS

The binary variable college plans (CP) is treated as the observed response and we study its dependence upon socioeconomic status (SES), intelligence (IQ) and parental encouragement (PE) treated as explanatory variables on an equal footing.

The data are input, without labels, from the layout of Table X.1; the labels are generated within the program.

A logistic regression model is fitted. The data are adequately explained by a model containing only main effects. The estimated parameters of Table X.3, *App. Stat.*, also the observed and fitted proportions of Table X.2, *App. Stat.*, are obtained.

Table X.1 Socioeconomic status, intelligence, parental encouragement and college plans for Wisconsin schoolboys

IQ	College plans	Parental encouragement	SES			
			L	LM	UM	H
L	Yes	Low	4	2	8	4
		High	13	27	47	39
	No	Low	349	232	166	48
		High	64	84	91	57
LM	Yes	Low	9	7	6	5
		High	33	64	74	123
	No	Low	207	201	120	47
		High	72	95	110	90
UM	Yes	Low	12	12	17	9
		High	38	93	148	224
	No	Low	126	115	92	41
		High	54	92	100	65
H	Yes	Low	10	17	6	8
		High	49	119	198	414
	No	Low	67	79	42	17
		High	43	59	73	54

PROGRAM

```
 2  JOB        'Example X'
 3
 4  OPEN       'XX'; 2
 5
 6  "Data:
-7      4   2   8   4
-8  ...
-9     43  59  73  54 : "
10
11  UNITS      [32]
12  TEXT       NX; !T(L,LM,UM,H)
13  &          NL; !T(Yes,No)
14  &          NE; !T(Low,High)
15  FACTOR     [LABELS=NX]  IQ,SES
16  &          [LABELS=NL]  Plans
17  &          [LABELS=NE]  Parents
18  TABLE      [CLASS=IQ,Plans,Parents,SES]  COUNTS
19  READ       [CHANNEL=2]  COUNTS

    Identifier  Minimum   Mean   Maximum  Values  Missing
       COUNTS     2.00   77.98   414.00      64        0    Skew
20
21  VARIATE    VCOUNT; !(#COUNTS)
22  TABLE      [CLASS=IQ,Parents,SES] Yes,No,Totals,Ratio,Fit
23  EQUATE     [OLDFORMAT=!((8,-8)4)]  COUNTS; Yes
24  EQUATE     [OLDFORMAT=!((-8,8)4)]  COUNTS; No
25
26  "We need binomial totals in order to fit logistic regression; and we also
```

```
-27   compute observed proportions for subsequent display:"
 28
 29   CALCULATE Totals = Yes+No
 30    &        Ratio = Yes/Totals
 31
 32   "Transfer table values to a variate and generate factor values."
 33
 34   EQUATE     Yes,Totals; YES,TOTALS
 35   GENERATE   IQ,Parents,SES
 36
 37   "The distribution is binomial, and the totals are required:"
 38
 39   MODEL      [DISTRIBUTION=BINOMIAL; LINK=LOGIT] YES; NBINOMIAL=TOTALS
 40   TERMS      IQ,Parents,SES
 41   FIT        IQ,Parents,SES
```

***** Regression Analysis *****

    Response variate: YES
     Binomial totals: TOTALS
        Distribution: Binomial
       Link function: Logit
        Fitted terms: Constant, IQ, Parents, SES

*** Summary of analysis ***
    Dispersion parameter is 1

	d.f.	deviance	mean deviance	deviance ratio
Regression	7	2237.37	319.624	303.97
Residual	24	25.24	1.051	
Total	31	2262.61	72.987	
Change	-7	-2237.37	319.624	303.97

* MESSAGE: The following units have large standardized residuals:
              32   2.22

*** Estimates of regression coefficients ***

	estimate	s.e.	t
Constant	-4.025	0.150	-26.82
IQ LM	0.594	0.124	4.81
IQ UM	1.333	0.119	11.16
IQ H	1.966	0.121	16.26
Parents High	2.455	0.101	24.22
SES LM	0.356	0.123	2.89
SES UM	0.662	0.120	5.54
SES H	1.414	0.121	11.69

* MESSAGE: s.e.s are based on dispersion parameter with value 1

```
 42                                               "Table X.3 App. Stat."
 43
 44   "Note the residual deviance 25.24 with 24 d.f. indicating satisfactory fit."
 45
 46   RKEEP      FITTED=FIT
 47   EQUATE     FIT; Fit
 48   CALCULATE  Fit = Fit/Totals
```

```
49  PRINT      [INTERLEAVE=3; NDOWN=3] Ratio,Fit;                        \
50             FIELD=10; DECIMALS=2                      "Table X.2 App. Stat."
```

			SES	L	LM	UM	H
IQ	Parents						
L	Low	Ratio		0.01	0.01	0.05	0.08
		Fit		0.02	0.02	0.03	0.07
	High	Ratio		0.17	0.24	0.34	0.41
		Fit		0.17	0.23	0.29	0.46
LM	Low	Ratio		0.04	0.03	0.05	0.10
		Fit		0.03	0.04	0.06	0.12
	High	Ratio		0.31	0.40	0.40	0.58
		Fit		0.27	0.35	0.42	0.61
UM	Low	Ratio		0.09	0.09	0.16	0.18
		Fit		0.06	0.09	0.12	0.22
	High	Ratio		0.41	0.50	0.60	0.78
		Fit		0.44	0.53	0.60	0.76
H	Low	Ratio		0.13	0.18	0.13	0.32
		Fit		0.11	0.15	0.20	0.34
	High	Ratio		0.53	0.67	0.73	0.88
		Fit		0.60	0.68	0.74	0.86

```
51  STOP
```

******** End of Example X. Maximum of 9280 data units used at line 40 (40080 left)

SUGGESTED FURTHER WORK

Fit a main-effects logistic model using scores (e.g. 1, 2, 3 and 4) to represent the categories of SES and IQ. Compare the fit of this model with that already fitted.

References

The abbreviation *App. Stat.* used frequently in this handbook refers to Cox and Snell (1981), referenced below.

Baxter, G. P. and Landstredt, O. W. (1940) A revision of the atomic weight of iodine. *J. Amer. Chem. Soc.*, **62**, 1829–34.

Biggers, J. D. and Heyner, S. (1961) Studies on the amino-acid requirements of cartilaginous long bone rudiments *in vitro. J. Exp. Zool.*, **147**, 95–112.

Brownlee, K. A.(1965) *Statistical Theory and Methodology in Science and Engineering* (2nd edn) Wiley, New York.

Cox, D. R. (1958) *Planning of Experiments*, Wiley, New York.

Cox, D. R. and Oakes, D. (1984) *Analysis of Survival Data*, Chapman and Hall, London.

Cox, D. R. and Snell, E. J. (1981) *Applied Statistics. Principles and examples*, Chapman and Hall, London.

Desmond, D. J. (1954) Quality control on the setting of voltage regulators. *Applied Statist.*, **3**, 65–73.

Digby, P. G. N. D., Galwey, N. W. and Lane, P. W. (1989) *Genstat 5: A Second Course*, Clarendon Press, Oxford.

Fedorov, V. D., Maximov, V. N. and Borgorov, V. G. (1968) Experimental development of nutritive media for micro-organisms. *Biometrika*, **55**, 43–51.

Fiegl, P. and Zelen, M. (1965) Estimation of exponential survival probabilities with concomitant information. *Biometrics*, **21**, 826–38.

Genstat 5 Committee (1987). *Genstat 5 Reference Manual*, Clarendon Press, Oxford.

Greenberg, R. A. and White, C. (1963) The sequence of sexes in human families. Paper presented to the 5th International Biometric Conference, Cambridge.

John, J. A. and Quenouille, M. H. (1977) *Experiments: Design and Analysis* (2nd edn), Griffin, London.

Johnson, N. L. (1967) Analysis of a factorial experiment. (Partially confounded 2^3). *Technometrics*, **9**, 167–70.

Lane, P. W., Galwey, N. W., and Alvey N. G. (1987) *Genstat 5: An Introduction,* Clarendon Press, Oxford.

Ling, R. F. (1984) Review of *Applied Statistics: Principles and examples.* Cox and Snell (1981). *J. Amer. Statis. Ass.*, **79**, 229–31.

MacGregor, G. A., Markandu, N. D., Roulston, J. E. and Jones, J. C. (1979) Essential hypertension: effect of an oral inhibitor of angiotension-converting enzyme. *Brit. Med. J.*, **2**, 1106–9.

Madsen, M. (1976) Statistical analysis of multiple contingency tables. Two examples. *Scand. J. Statist.*, **3**, 97–106.

Mooz, W. E. (1978) Cost analysis of light water reactor power plants. *Report R-2304-DOE.* Rand Corp., Santa Monica, CA.

Morton, A. Q. (1965) The authorship of Greek prose (with discussion).*J. R. Statist. Soc.*, A, **128**, 169–233.

Pearson, E. S. and Hartley, H. O. (1966) *Biometrika Tables for Statisticians*, Vol. 1 (3rd edn), Cambridge University Press, Cambridge.

Proschan, F. (1963) Theoretical explanation of observed decreasing failure rate. *Technometrics*, **5**, 375–83.

Ries, P. N. and Smith, H. (1963) The use of chi-square for preference testing in multidimensonal problems. *Chem. Eng. Progress,* **59**, 39–43.

Sewell, W. H. and Shah, V. P. (1968) Social class, parental encouragement and educational aspirations. *Amer. J. Sociol.*, **73**, 559–72.

Woolf, B. (1955) On estimating the relation between blood group and disease. *Ann. Hum. Genetics*, **19**, 251–3.

Index

Genstat command names are given in capitals, and indexed alphabetically. (... indicates that the command occurs frequently; references after the first three instances in the Examples are not listed.) Features of Genstat are indexed under the heading 'Genstat language'.